Boundary Element Techniques in Engineering

Boundary Element Techniques in Engineering

C. A. BREBBIA, DIPL. ING., PH.D.

Senior Lecturer in Civil Engineering
University of Southampton, UK

S. WALKER M.A. (CANTAB), M.SC.

Consultant Engineer,
Structural Dynamics Ltd, Southampton, UK

NEWNES–BUTTERWORTHS

LONDON · BOSTON
Sydney · Wellington · Durban · Toronto

The Butterworth Group

United Kingdom **Butterworth & Co. (Publishers) Ltd**
London: 88 Kingsway, WC2B 6AB

Australia **Butterworths Pty Ltd**
Sydney: 586 Pacific Highway, Chatswood, NSW 2067
Also at Melbourne, Brisbane, Adelaide and Perth

Canada **Butterworth & Co. (Canada) Ltd**
Toronto: 2265 Midland Avenue, Scarborough, Ontario MIP 4S1

New Zealand **Butterworths of New Zealand Ltd**
Wellington: T & W Young Building, 77–85 Customhouse Quay, 1,
CPO Box 472

South Africa **Butterworth & Co. (South Africa) (Pty) Ltd**
Durban: 152–154 Gale Street

USA **Butterworth (Publishers) Inc.**
Boston: 10 Tower Office Park, Woburn, Mass. 01801

First published 1980

British Library Cataloguing in Publication Data

Brebbia, Carlos Alberto
Boundary element techniques in engineering.
1. Engineering mathematics
2. Boundary value problems
I. Title II. Walker, Stephen, b. 1949
620′.001′515353 TA347.B/ 79–4272

ISBN 0–408–00340–5

Typeset by The Macmillan Co. of India Ltd., Bangalore
Printed in England by The Camelot Press Ltd., Southampton

Preface

The boundary element method originated at Southampton University from previous work on classical integral equations and finite elements. The new method has the advantages of both techniques, that is, it reduces the dimensions of the problem by one as boundary integral equations and it allows for complex surface elements to define the external surface of the domain. It represents an advance over classical finite elements and overcomes many of the main disadvantages such as the difficulty of defining with accuracy domains extending to infinity, having to solve large systems of equations and the need to define complicated data structure.

The method is ideally suited to the solution of many two- and three-dimensional problems in elasticity and potential theory for which finite elements, although popular, are inefficient. In the authors' opinion, finite elements are appropriate for problems such as shells, highly anisotropic media, some non-linear and convective problems but their applicability to a wide range of problems has been overstated. This is doubtless due to the energy of finite element researchers as well as to the coincidental appearance of the method with the development of the first generation of powerful computers. The authors are also aware that the boundary element may not appeal to present-day finite element-oriented engineers and consequently dedicate this book to the coming generation with the warning that they should never take any written work as definitive. Therefore we not only expect but clearly hope that this book will be superseded in due course and that, when this happens, we shall be able to bow gracefully rather than enter into fruitless argument.

It is suggested that readers who wish to have an introduction to boundary elements and a basic knowledge of how boundary element programs are developed should refer to the main author's earlier book, *The Boundary Element Method for Engineers* (Pentech Press, London, 1978).

Contents

1 Approximate Methods

1.1 INTRODUCTION

Once the mathematical equations governing a problem have been established, then excepting very simple cases, they must be solved using an approximate method. The type of approximation used is important as it affects the accuracy of the results and the economy of the solution. Unfortunately it is difficult to compare different approximating techniques as they are usually presented as totally unrelated. The finite element method for instance, is viewed as a virtual work method, a variational technique or a weighted residual method, depending on the particular application under consideration or the researcher's preference. Boundary solutions are sometimes presented as a consequence of the reciprocity principles and only recently they have been interpreted as weighted residual solutions. In what follows we will present a rational classification of the different approximate methods based on the weighted residual formulation and on the type of approximating and weighting functions used.

Weighted residual formulations are more general than classical variational principles which can only be applied to a restricted type of operator. Weighted residuals can be used with complex non-self-adjoint operators equally well as with self-adjoint ones. Variational techniques, finite differences, finite elements, integral type formulations and many other methods can be interpreted as special cases of weighted residual methods. The converse is not generally true.

The aim of an approximate solution is to reduce a governing equation (or set of equations) plus boundary conditions to a system of algebraic equations. This is usually done by subdividing the continuum into a number of cells or elements and assuming over each of these a known variation of the approximating and weighting functions. Consider for instance, an equation such as Laplace's for simplicity. This equation can be written as,

$$\nabla^2 u_0 = 0 \qquad \text{in } \Omega \tag{1.1}$$

where u_0 indicates the exact solution and Ω is the domain under consideration. The corresponding boundary conditions may be of the following two types:

$$\begin{aligned} &\text{conditions } u_0 = \bar{u} && \text{on } \Gamma_1 \\ &\text{conditions } q_0 = \partial u_0/\partial n = \bar{q} && \text{on } \Gamma_2 \end{aligned} \tag{1.2}$$

where Γ_1 and Γ_2 are parts of the boundary where one or the other condition applies. The total boundary is $\Gamma = \Gamma_1 + \Gamma_2$.

The exact solution u_0 can be found only for a few and simple cases and generally the solution will have to be approximated. This can be done by using a set of known linearly independent functions ϕ_i and unknown coefficients α_i or u_i. (u_i are values of the u functions at certain points or nodes and α_i generalised coefficients that are not nodal values.) In general we will prefer to use the nodal values as they have a clear physical meaning but when the nodes are not defined we will use the α's.

Let us write the approximation for u as

$$u = \alpha_1 \phi_1 + \alpha_2 \phi_2 + \ldots \tag{1.3}$$

Introducing this function into (1.1) and (1.2) will produce an error (excepting the case for which (1.3) is the exact solution, i.e. $u = u_0$), such that,

$$\nabla^2 u \neq 0 \qquad \text{in } \Omega \tag{1.4}$$

and on the boundary

$$\begin{aligned} u - \bar{u} \neq 0 \qquad & \text{on } \Gamma_1 \\ q - \bar{q} \neq 0 \qquad & \text{on } \Gamma_2 \end{aligned} \tag{1.5}$$

$q = \partial u/\partial n$.

We can now define error functions in the domain and on the boundary such that,

$$\varepsilon = \nabla^2 u \neq 0 \qquad \text{in } \Omega \tag{1.6}$$

and,

$$\begin{aligned} \varepsilon_1 &= u - \bar{u} \neq 0 \qquad \text{on } \Gamma_1 \\ \varepsilon_2 &= q - \bar{q} \neq 0 \qquad \text{on } \Gamma_2 \end{aligned} \tag{1.7}$$

Our aim is now to make this error as small as possible over the domain and on the boundary. In order to do so the errors can be distributed and the way in which this distribution is carried out produces different types of *weighted residual techniques.*

1.2 WEIGHTED RESIDUAL TECHNIQUES

The simplest of these techniques starts by exactly satisfying the boundary conditions (i.e. $\varepsilon_1 = \varepsilon_2 = 0$) and distributing the error according with a weighting function w. This function is such that it identically satisfies the homogeneous boundary conditions and can be written as,

$$w = \beta_1 \psi_1 + \beta_2 \psi_2 + \beta_3 \psi_3 + \ \ldots \tag{1.8}$$

The ψ_i are members of a set of known linearly independent functions and β_i are arbitrary coefficients. We will call them w_i when they are arbitrary *nodal* coefficients.

The distribution of the error function ε can now be carried out by multiplying it by the weighting function w and integrating over the domain, i.e.

$$\int_{\Omega} \varepsilon w \, \mathrm{d}\Omega = \int_{\Omega} (\nabla^2 u) w \, \mathrm{d}\Omega = 0 \tag{1.9}$$

In this way the error is distributed in accordance with the functions in w. Note that because the β_i coefficients are arbitrary, (1.9) can also be written as,

$$\int_{\Omega} (\nabla^2 u) \psi_i \, \mathrm{d}\Omega = 0 \quad \text{for } i = 1, 2, \ldots, n \tag{1.10}$$

Classical finite difference techniques for instance can be interpreted as a special case of equation (1.9) for which the weighting functions are *Dirac* delta functions, and the β_i are nodal arbitrary coefficients (w_i), such that,

$$w = w_1 \delta_1 + w_2 \delta_2 + w_3 \delta_3 + \ \ldots \tag{1.11}$$

The Dirac delta function is a generalised function which can be defined as the limit of a normal function such as

$$\lim_{N \to \infty} \frac{\sin[N(x - x_i)]}{\pi(x - x_i)} = \delta(x - x_i) \tag{1.12}$$

In the limit this function is zero at every point except where the argument is zero where it is infinite (i.e. at $x = x_i$). Thus it represents a point singularity at the 'source' point x_i.

The delta function has the property that

$$\int_{-\infty}^{\infty} \delta(x - x_i)\,\mathrm{d}x = \int_{x_i - a}^{x_i + a} \delta(x - x_i)\,\mathrm{d}x = 1 \tag{1.13}$$

for a any positive number.

We also have that for any function $f(x)$ (continuous at x_i) we can write

$$\int_{-\infty}^{\infty} f(x)\delta(x - x_i)\,\mathrm{d}x = \int_{x_i - a}^{x_i + a} f(x)\delta(x - x_i)\,\mathrm{d}x = f(x_i) \tag{1.14}$$

Where it is not necessary to represent the arguments of the functions explicitly in the analysis we shall use the shorthand notation putting

$$\delta(x - x_i) = \delta_i$$

and

$$f(x_i) = f_i \tag{1.15}$$

for simplicity. Where it is important to include the arguments for clarity of explanation we shall use the full notation writing x and x_i explicitly.

The approximating functions are given by

$$u = u_1\phi_1 + u_2\phi_2 + u_3\phi_3 + \ldots \tag{1.16}$$

(The ϕ_i are localised second-order polynomials taken over the subdomain surrounding the node for the central difference scheme.)

Substituting (1.16) into (1.9) produces the following equation at each node 'i':

$$\int_{\Omega} \varepsilon\delta_i\,\mathrm{d}\Omega = \varepsilon_i = 0 \tag{1.17}$$

The above presentation gives the same numerical results as the central finite difference scheme for quadratic approximating functions. The procedure is however, more general as it allows us equally well to use other types of ϕ_i functions. Furthermore the basic ideas can be generalised. Using curvilinear coordinates these generalisations give rise to the 'cell collocation' method.[1] Basically this is a subregion collocation technique as shown in Example 1.1.

Example 1.1

Consider the equation

$$\frac{d^2u}{dx^2}+u+x=0 \qquad x\in[0,\,1] \tag{a}$$

where x varies from 0 to 1 with boundary conditions

$$u(0)=u(1)=0 \tag{b}$$

Figure 1.1 Point under consideration

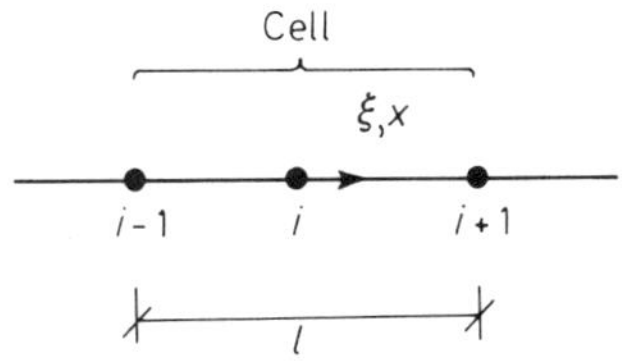

We can now propose over each subregion or 'cell' (Figure 1.1) the following approximating function:

$$u=u_{i-1}\phi_1+u_i\phi_2+u_{i+1}\phi_3 \tag{c}$$

where,

$$\begin{aligned}\phi_1&=\tfrac{1}{2}\xi(\xi-1)\\ \phi_2&=(1-\xi)(1+\xi)\\ \phi_3&=\tfrac{1}{2}\xi(1+\xi)\end{aligned} \tag{d}$$

ξ is a dimensionless coordinate, $\xi=x/\tfrac{1}{2}l$. If we differentiate equation (a) with functions (d) one finds the following algebraic relationship:

$$\begin{aligned}\varepsilon_i&=\left(\frac{d^2u}{dx^2}+u+x\right)_i\\ &=\left(\frac{4}{l^2}u_{i-1}-\frac{8}{l^2}u_i+\frac{4}{l^2}u_{i+1}\right)+u_i+x_i\\ &=\frac{4}{l^2}(u_{i-1}-2u_i+u_{i+1})+u_i+x_i\end{aligned} \tag{e}$$

This ε_i is made equal to zero at i according to equation (1.11), which gave the same result as central finite differences, for this particular case.

METHOD OF MOMENTS

Equation (1.9) can be used to generate a wide variety of weighted residual techniques such as the method of moments, the original Galerkin method, collocation, subregions, etc.

The *method of moments* consists in taking moments of the error function. For instance for a one-dimensional problem this implies taking moments of ε with respect to the following function:

$$1, x, x^2, \ldots \tag{1.18}$$

These are the ψ_i functions of equation (1.8). Hence we can write a weighting function such that,

$$\begin{aligned} w &= \beta_1\psi_1 + \beta_2\psi_2 + \beta_3\psi_3 + \ldots \\ &= \beta_1 1 + \beta_2 x + \beta_3 x^2 + \ldots \end{aligned} \tag{1.19}$$

The method is illustrated with the following example.

Example 1.2

Assume to have the same equation and boundary conditions as seen in Example 1.1. The error function is

$$\varepsilon = \frac{d^2u}{dx^2} + u + x \tag{a}$$

We can propose an approximating function that satisfies the boundary conditions and such as,

$$u = \alpha_1\phi_1 + \alpha_2\phi_2 + \ldots = \alpha_1 x(1-x) + \alpha_2 x^2(1-x) + \ldots \tag{b}$$

Note that the $\phi_1, \phi_2, \ldots$ functions satisfy the homogeneous boundary conditions at $x = 0$ and $x = 1$. The α_i coefficients are unknown and are not directly associated with *nodal* values of the u functions as we have not defined any nodes over the domain. The error function for two α_i coefficients results

$$\varepsilon = x + (-2 + x - x^2)\alpha_1 + (2 - 6x + x^2 - x^3)\alpha_2 \qquad \text{(c)}$$

This function needs to be orthogonalised with respect to 1 and x, i.e. we take the following two moments:

$$\int_0^1 \varepsilon\, 1\, \mathrm{d}x = 0; \qquad \int_0^1 \varepsilon x\, \mathrm{d}x = 0 \qquad \text{(d)}$$

Integrating equations (d) we find a system of 2×2 equations from where the α_1, α_2 coefficients can be found. The system is

$$\begin{bmatrix} \frac{11}{6} & \frac{11}{12} \\ \frac{11}{22} & \frac{19}{20} \end{bmatrix} \begin{Bmatrix} \alpha_1 \\ \alpha_2 \end{Bmatrix} = \begin{Bmatrix} \frac{1}{2} \\ \frac{1}{3} \end{Bmatrix} \qquad \text{(e)}$$

The solution of these equations gives,

$$\alpha_1 = \tfrac{122}{649}, \quad \alpha_2 = \tfrac{110}{649} \qquad \text{(f)}$$

Hence the approximate solution becomes,

$$u = x(1 - x)\left(\tfrac{122}{649} + \tfrac{110}{649}x\right) \qquad \text{(g)}$$

which compares favourably with the exact one,

$$u_e = \frac{\sin x}{\sin 1} - x \qquad \text{(h)}$$

GALERKIN METHOD

The original Galerkin method is also represented by equation (1.9). Here we start with an approximating function,

$$u = \alpha_1 \phi_1 + \alpha_2 \phi_2 + \alpha_3 \phi_3 + \ldots \qquad (1.20)$$

and choose the weighting function w such that,

$$w = \beta_1 \phi_1 + \beta_2 \phi_2 + \beta_3 \phi_3 + \ldots \qquad (1.21)$$

that is the same functions are used for u and w. The weighted residual statement,

$$\int_\Omega \varepsilon w\, \mathrm{d}\Omega = 0 \qquad (1.22)$$

can now be written as

$$\int_{\Omega} \varepsilon \phi_i \, \mathrm{d}\Omega = 0 \quad i = 1, 2, \ldots \tag{1.23}$$

It is also common in Galerkin method to write the w function as Δu, i.e.

$$w = \Delta u$$

where

$$\Delta u = \Delta\alpha_1 \phi_1 + \Delta\alpha_2 \phi_2 + \Delta\alpha_3 \phi_3 + \ldots \tag{1.24}$$

and $\Delta\alpha_i = \beta_i$.

This representation is sometimes preferred to indicate that w can be identified with a variation or virtual quantity (such as virtual displacements or velocities). Expression (1.24) also indicates that the same functions are used for u as for Δu. This property is important as it produces symmetry in the expressions as we will see in Example 1.3.

Example 1.3

Let us return to our original equation, defined in Example 1.1 and try to solve it using Galerkin's. We can choose the same approximating functions as in Example 1.2, i.e.

$$u = \alpha_1 \phi_1 + \alpha_2 \phi_2 = \alpha_1 x(1-x) + \alpha_2 x^2 (1-x) \tag{a}$$

Note that as previously, α_1, α_2 are not nodal values of u but unknown generalised coefficients.

The weighted residual statement is,

$$\int_x \varepsilon w \, \mathrm{d}x = 0 \tag{b}$$

which produces the following two equations:

$$\int_x \varepsilon \phi_1 \, \mathrm{d}x = 0 \quad \text{and} \quad \int_x \varepsilon \phi_2 \, \mathrm{d}x = 0 \tag{c}$$

The ε function is the same as in Example 1.2. Hence,

$$\int_x [x+(-2+x-x^2)\alpha_1+(2-6x+x^2-x^3)\alpha_2]\,[x(1-x)]\,\mathrm{d}x = 0$$

$$\int_x [x+(-2+x-x^2)\alpha_1+(2-6x+x^2-x^3)\alpha_2]\,[x^2(1-x)]\,\mathrm{d}x = 0 \tag{d}$$

which gives the following system after integration:

$$\begin{bmatrix} \frac{3}{10} & \frac{3}{20} \\ \frac{3}{20} & \frac{13}{105} \end{bmatrix} \begin{Bmatrix} \alpha_1 \\ \alpha_2 \end{Bmatrix} = \begin{Bmatrix} \frac{1}{12} \\ \frac{1}{20} \end{Bmatrix} \tag{e}$$

Note that the matrix is symmetric because the approximate and weighting functions are the same.

The α's are

$$\alpha_1 = \tfrac{71}{369}, \qquad \alpha_2 = \tfrac{7}{41} \tag{f}$$

The approximate solution is,

$$u = x(1-x)\left(\tfrac{71}{369}+\tfrac{7}{41}x\right) \tag{g}$$

which compares well with the exact solution.

Example 1.4

The same Example 1.3 can be solved using trigonometric functions as approximating solutions. Trigonometric functions are useful because, for linear problems, they give a set of uncoupled equations instead of the fully populated matrix shown in the previous example.

Let us consider the same equation as previously and take

$$u = \alpha_1 \sin(\pi x) + \alpha_2 \sin(2\pi x) \tag{a}$$

Hence

$$\varepsilon = \frac{\mathrm{d}^2 u}{\mathrm{d}x^2} + u + x = -\pi^2[\alpha_1 \sin(\pi x) + 4\alpha_2 \sin(2\pi x)] + [\alpha_1 \sin(\pi x) + \alpha_2 \sin(2\pi x)] + x \tag{b}$$

The integrals now are,

$$\int_0^1 \varepsilon \sin(\pi x)\,\mathrm{d}x = 0 \qquad \int_0^1 \varepsilon \sin(2\pi x)\,\mathrm{d}x = 0 \tag{c}$$

Taking orthogonality into consideration, i.e.

$$\int_0^1 \sin(m\pi x)\sin(n\pi x)\,\mathrm{d}x = \begin{cases} 0 & \text{for } m \neq n \\ \frac{1}{2} & \text{for } m = n. \end{cases} \qquad \text{(d)}$$

we obtain two equations,

$$\left(\frac{1}{2}-\frac{\pi^2}{2}\right)\alpha_1 + \frac{1}{\pi} = 0 \qquad \left(\frac{1}{2}-\frac{4\pi^2}{2}\right)\alpha_2 - \frac{1}{2\pi} = 0$$

Hence

$$\alpha_1 = \frac{2}{\pi}\frac{1}{(\pi^2-1)} = 0{\cdot}07177$$

$$\alpha_2 = \frac{2}{2\pi}\frac{1}{(1-4\pi^2)} = -0{\cdot}00827$$

The value of the u function is

$$u = \alpha_1 \sin(\pi x) + \alpha_2 \sin(2\pi x)$$

The value of u at the centre ($x = \frac{1}{2}$) is approximately the same as in the previous methods. Notice that for this case

$$u_c = \alpha_1$$

1.3 WEAK FORMULATIONS

We have assumed up to now that all boundary conditions were satisfied on Γ_1 and Γ_2. In many cases however, it is more convenient to satisfy the conditions on Γ_1 identically ($u = \bar{u}$ for Laplace's) but to approximate those on Γ_2 ($q = \bar{q}$ for Laplace's). The boundary error function ε_2 produced in this way can also be weighted in a similar form as the ε error in the domain was weighted, i.e.

$$\int_\Omega (\nabla^2 u) w \,\mathrm{d}\Omega = \int_{\Gamma_2} (q - \bar{q}) w \,\mathrm{d}\Gamma \qquad (1.25)$$

or

$$\int_\Omega \varepsilon\, w \,\mathrm{d}\Omega = \int_{\Gamma_2} \varepsilon_2\, w \,\mathrm{d}\Gamma \qquad (1.26)$$

This formulation requires approximating functions that only satisfy the conditions on Γ_1 and weighting functions which only fulfil the homogeneous conditions on Γ_2. The sign of the right-hand side term can be deduced by integrating by parts as will be shown.

Furthermore we can reduce the order required for the approximating function by integrating the left-hand side of equation (1.27) by parts. This gives,

$$\int_{\Omega} \frac{\partial u}{\partial x_k} \frac{\partial w}{\partial x_k} \, \mathrm{d}\Omega = \int_{\Gamma_2} \bar{q} \, w \, \mathrm{d}\Gamma \tag{1.27}$$

where the indicial notation has been used, with Einstein summation convention.

Note that the u functions in equation (1.25) needed to be continuous up to their second-order derivatives, while the w functions were continuous (but their derivatives need not be continuous). Now (equation (27)) the order of continuity for u and w is the same, i.e. continuity up to their first derivatives.

Formulations of the type of equation (1.27) are called 'weak' formulations and are the starting point for the finite element method. For this case the u and w are defined as,

$$u = u_1 \phi_1 + u_2 \phi_2 + u_3 \phi_3 + \ \dots \tag{1.28}$$

$$w = \Delta u = \Delta u_1 \phi_1 + \Delta u_2 \phi_2 + \Delta u_3 \phi_3 + \ \dots \tag{1.29}$$

where u_i are nodal unknowns. Hence equation (1.27) is usually written,

$$\int_{\Omega} \frac{\partial u}{\partial x_k} \frac{\partial \Delta u}{\partial x_k} \, \mathrm{d}\Omega = \int_{\Gamma_2} \bar{q} \, \Delta u \, \mathrm{d}\Gamma \tag{1.30}$$

This procedure has the advantage that the resulting matrices are symmetric.

Example 1.5

Consider again the second-order equation,

$$\frac{\mathrm{d}^2 u_0}{\mathrm{d}x^2} + u_0 + x = 0 \tag{a}$$

with boundary conditions $u_0 = 0$ at $x = 0$ but at $x = 1$ we impose the condition

$$q = \mathrm{d}u_0/\mathrm{d}x = \bar{q} \tag{b}$$

where $\bar{q}$ is known.

Let us start with the following approximation:

$$u = \alpha_1 + \alpha_2 x + \alpha_3 x^2 \tag{c}$$

and try to satisfy the *two boundary conditions*

$$\begin{aligned} &\text{at } x = 0 \qquad u = \alpha_1 = 0 \\ &\text{at } x = 1 \qquad q = \mathrm{d}u/\mathrm{d}x = 2\alpha_3 + \alpha_2 = \bar{q} \end{aligned} \tag{d}$$

Hence,

$$\alpha_2 = \bar{q} - 2\alpha_3 \tag{e}$$

Calling $\alpha_3 = \alpha$ we can rewrite (c) as,

$$u = \alpha x^2 + (\bar{q} - 2\alpha)x = \bar{q}x - 2\alpha x + \alpha x^2 \tag{f}$$

The residual for this case is

$$\varepsilon = \frac{\mathrm{d}^2 u}{\mathrm{d}x^2} + u + x = 2\alpha + \bar{q}x - 2\alpha x + \alpha x^2 + x \tag{g}$$

This error can be distributed by orthogonalising it with respect to the function $(x^2 - 2x)$ only, i.e.

$$w = \Delta u = \Delta\alpha(x^2 - 2x) \tag{h}$$

This gives the following weighted residual statement:

$$\int_0^1 (2\alpha + \bar{q}x - 2\alpha x + \alpha x^2 + x)\,(x^2 - 2x)\,\mathrm{d}x = 0$$

$$\tfrac{24}{5}\alpha + \tfrac{5}{2}\bar{q} + \tfrac{5}{2} = 0 \tag{i}$$

$$\alpha = -0{\cdot}5208(\bar{q} + 1)$$

The value of u is given by

$$u = \bar{q}x - 0{\cdot}5208(\bar{q} + 1)(x^2 - 2x) \tag{j}$$

This result has been obtained satisfying both boundary conditions. Let us now consider what happens if the first condition ($u = 0$) is satisfied but the other ($q = \bar{q}$) is approximated. Now we have,

$$\int_0^1 \left(\frac{d^2u}{dx^2} + u + x\right) \Delta u \, dx = \left[\left(\frac{du}{dx} - \bar{q}\right) \Delta u\right]_{\text{at } x=1} \tag{k}$$

Integrating the first term by parts one obtains,

$$\int_0^1 \left(\frac{du}{dx}\frac{d\Delta u}{dx} - (u+x)\Delta u\right) dx = [\bar{q}\,\Delta u]_{\text{at } x=1} \tag{l}$$

We can take the following approximation:

$$u = \alpha_1 x^2 + \alpha_2 x \tag{m}$$

which identically satisfies $u = 0$ at $x = 0$, but not $q = \bar{q}$ at $x = 1$. Integral (l) becomes,

$$\begin{aligned}\int_0^1 & [(2\alpha_1 x + \alpha_2)(2\Delta\alpha_1 x + \Delta\alpha_2) - (\alpha_1 x^2 + \alpha_2 x + x)\times \\ & \times (\Delta\alpha_1 x^2 + \Delta\alpha_2 x)]\, dx \\ & = \bar{q}(\Delta\alpha_1 + \Delta\alpha_2).\end{aligned} \tag{n}$$

After integration one finds the following system:

$$\begin{bmatrix} \frac{17}{15} & \frac{3}{4} \\ \frac{3}{4} & \frac{2}{3} \end{bmatrix} \begin{Bmatrix} \alpha_1 \\ \alpha_2 \end{Bmatrix} = \begin{Bmatrix} \bar{q} + \frac{1}{4} \\ \bar{q} + \frac{1}{3} \end{Bmatrix} \tag{o}$$

This system gives,

$$\begin{aligned} \alpha_1 &= -0{\cdot}4319 - 0{\cdot}4322\bar{q} \\ \alpha_2 &= 0{\cdot}9859 + 1{\cdot}9864\bar{q} \end{aligned} \tag{p}$$

Let us now compute the values of q at $x = 1$, i.e.

$$q = \left.\frac{du}{dx}\right|_{x=1} = 2\alpha_1 + \alpha_2 = 0{\cdot}1221 + 1{\cdot}220\bar{q} \tag{q}$$

Hence the value of $\bar{q}$ at $x = 1$ is not equal to $\bar{q}$ but is an approximation.

At the centre point $x = \frac{1}{2}$ we have,

$$u_c = \tfrac{1}{4}\alpha_1 + \tfrac{1}{2}\alpha_2 = 0{\cdot}3850 + 0{\cdot}8852\bar{q} \tag{r}$$

Note that the first solution (equation (j)) gives,

$$u_c = 0{\cdot}3906 + 0{\cdot}8906\bar{q} \tag{s}$$

1.4 THE INVERSE PROBLEM

Sometimes it is convenient to take into account the conditions on Γ_1 ($u = \bar{u}$ for the Laplace equation) in the weighted residual statement. This produces an approximation for these conditions and the error integrals can now be written as follows,

$$\int_{\Omega} \varepsilon\, w\, \mathrm{d}\Omega = \int_{\Gamma_2} \varepsilon_2\, w\, \mathrm{d}\Gamma - \int_{\Gamma_1} \varepsilon_1 \frac{\partial w}{\partial n} \mathrm{d}\Gamma \tag{1.31}$$

or

$$\int_{\Omega} (\nabla^2 u)\, w\, \mathrm{d}\Omega = \int_{\Gamma_2} (q-\bar{q})\, w\, \mathrm{d}\Gamma - \int_{\Gamma_1} (u-\bar{u}) \frac{\partial w}{\partial n} \mathrm{d}\Gamma \tag{1.32}$$

The inclusion of the last integral with the weighting function $\partial w/\partial n$ can be justified in different ways but it is simpler to accept (1.32) on dimensional grounds and to prove that the minus sign is correct by integrating by parts. Integrating (1.32) by parts once, we obtain

$$\int_{\Omega} \frac{\partial u}{\partial x_k} \frac{\partial w}{\partial x_k} \mathrm{d}\Omega = \int_{\Gamma_2} \bar{q}\, w\, \mathrm{d}\Gamma + \int_{\Gamma_1} q\, w\, \mathrm{d}\Gamma - \int_{\Gamma_1} \bar{u} \frac{\partial w}{\partial n} \mathrm{d}\Gamma + \int_{\Gamma_1} u \frac{\partial w}{\partial n} \mathrm{d}\Gamma \tag{1.33}$$

This expression is seldom used as the boundary conditions on Γ_1 are usually easy to satisfy when working with domain methods such as finite elements. In this case the three integrals on Γ_1 disappear.

Integrating equation (1.33) again by parts we obtain the following statement:

$$\int_{\Omega} (\nabla^2 w)\, u\, \mathrm{d}\Omega = -\int_{\Gamma_2} \bar{q}\, w\, \mathrm{d}\Gamma - \int_{\Gamma_1} q\, w\, \mathrm{d}\Gamma + \int_{\Gamma_2} u \frac{\partial w}{\partial n} \mathrm{d}\Gamma + \int_{\Gamma_1} \bar{u} \frac{\partial w}{\partial n} \mathrm{d}\Gamma \tag{1.34}$$

Formula (1.34) is the inverse of the original expression (1.32) and it offers the possibility of working only on the boundary if a solution is found for w such that $\nabla^2 w \equiv 0$. Note that now the weighting function w needs to be continuous up to the second derivatives but u only needs to have continuity of the function itself (but not of the derivatives).

Here again the weighting function w can be found using the same functions as for the approximating function u or different functions.

Statements (1.32), (1.33) and (1.34) permit us to establish a classification of the different approximating techniques in accordance with the type of statement one uses and the relationship between the approximating and the weighting functions. Note that the first two statements usually give the domain methods but the last one can be used to build boundary solutions.

Figure 1.2 illustrates the different types of approximating techniques some of which have already been explained.

Example 1.6

Consider again the equation,

$$\frac{\mathrm{d}^2 u}{\mathrm{d}x^2} + u + x = 0 \qquad \text{in } \Omega \tag{a}$$

with boundary conditions $u = 0$ at $x = 0$ and at $x = 1$.

Statement (32) can be written (note all the boundary is of the Γ_1 type),

$$\int_0^1 \left(\frac{\mathrm{d}^2 u}{\mathrm{d}x^2} + u + x\right) w \, \mathrm{d}x + \left[(u - \bar{u}) \frac{\mathrm{d}w}{\mathrm{d}x}\right]_{x=0}^{x=1} = 0 \tag{b}$$

Integrating by parts twice we obtain

$$\int_0^1 \left(\frac{\mathrm{d}^2 w}{\mathrm{d}x^2} + w\right) u \, \mathrm{d}x + \int_0^1 x w \, \mathrm{d}x + [q w]_{x=0}^{x=1} - \left[\bar{u} \frac{\mathrm{d}w}{\mathrm{d}x}\right]_0^1 = 0 \tag{c}$$

Note that now we need to find a solution that satisfies the equation

$$\frac{\mathrm{d}^2 w}{\mathrm{d}x^2} + w = 0 \tag{d}$$

The general solution is,

$$w = \beta_1 \cos x + \beta_2 \sin x \tag{e}$$

with

$$\mathrm{d}w/\mathrm{d}x = -\beta_1 \sin x + \beta_2 \cos x \tag{f}$$

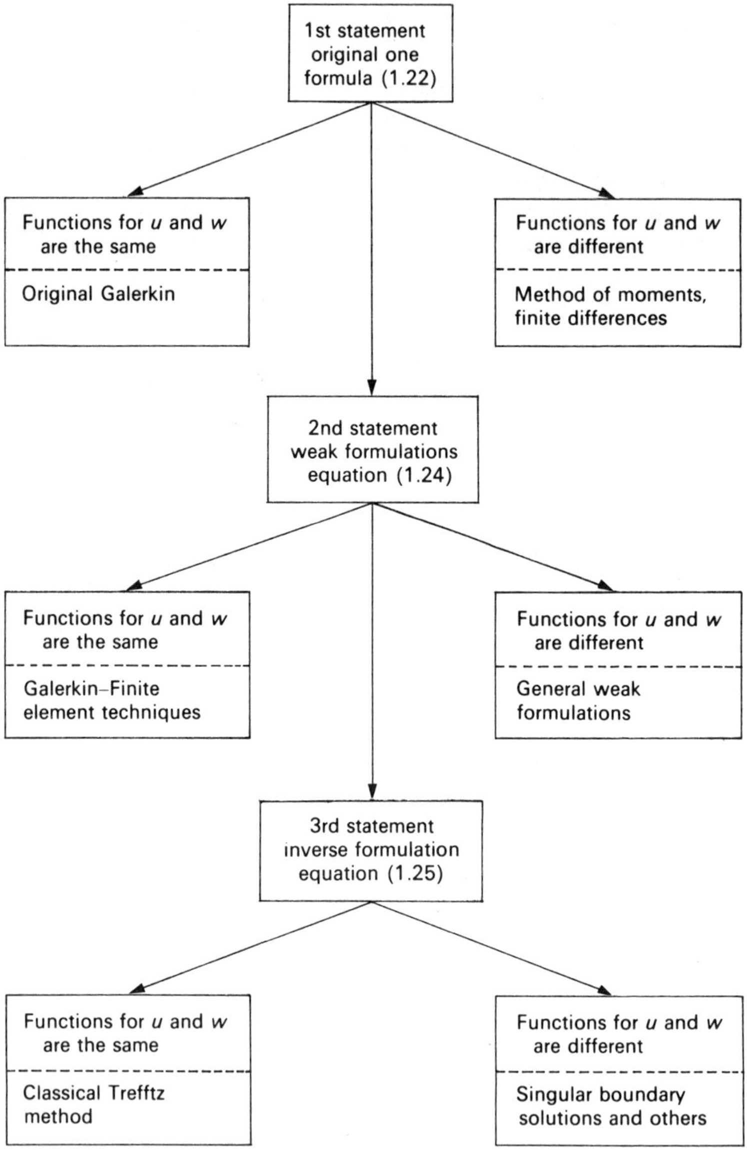

Figure 1.2 Different approximating techniques

Note that the solution for u and w can be of any form and that these functions are defined only on the boundary. In this particular example the boundary reduces to two points, hence we have as unknowns, q_0 and q_1 (i.e. $\mathrm{d}u/\mathrm{d}x$ at 1 and 0).

Consider now equation (c) and substitute into it the weighting function (e). This gives,

$$\int_0^1 x\,w\,\mathrm{d}x + [q\,w]_1 - [q\,w]_0 = 0$$

$$\int_0^1 x(\beta_1 \cos x + \beta_2 \sin x)\,\mathrm{d}x + q_1(\beta_1 \cos 1 + \beta_2 \sin 1) - q_0(\beta_1) = 0$$

This gives the following equations (note that β_1 and β_2 are arbitrary):

$$\begin{aligned} \int_0^1 x \cos x\,\mathrm{d}x &= -(q_1 \cos 1 - q_0) \\ \int_0^1 x \sin x\,\mathrm{d}x &= -q_1 \sin 1 \end{aligned} \qquad \text{(h)}$$

$$q_1 = \frac{\cos 1}{\sin 1} - 1, \qquad q_0 = \frac{1}{\sin 1} - 1 \qquad \text{(i)}$$

These are exact boundary fluxes $\partial u/\partial x$ at $x = 1$ and $\partial u/\partial x$ at $x = 0$

1.5 BOUNDARY METHODS

The last example showed an application of the inverse relationship to find a boundary solution. Let us now try to use the integrals seen in Section 1.3 to establish boundary solutions. Assume for instance, that we have a w function such that $\nabla^2 w = 0$. Hence formula (1.34) becomes,

$$\int_{\Gamma_2} \bar{q}\,w\,\mathrm{d}\Gamma + \int_{\Gamma_1} q\,w\,\mathrm{d}\Gamma = \int_{\Gamma_2} u \frac{\partial w}{\partial n}\,\mathrm{d}\Gamma + \int_{\Gamma_1} \bar{u} \frac{\partial w}{\partial n}\,\mathrm{d}\Gamma \qquad (1.35)$$

This relationship is the starting point for the 'direct' boundary element method.

If the same functions are used for w as for the approximate function u, one can write,

$$\int_{\Gamma_2} \bar{q}\, \Delta u \,\mathrm{d}\Gamma + \int_{\Gamma_1} q\, \Delta u \,\mathrm{d}\Gamma = \int_{\Gamma_2} u \frac{\partial \Delta u}{\partial n} \mathrm{d}\Gamma + \int_{\Gamma_1} \bar{u} \frac{\partial \Delta u}{\partial n} \mathrm{d}\Gamma \quad (1.36)$$

It is interesting to note that in this case we do not need to use the inverse relationship (equation (1.36)), because u as well as Δu satisfies the governing equations. Thus we can start with equation (1.33) for which now $\nabla^2 u = 0$ and obtain,

$$\int_{\Gamma_2} (q - \bar{q})\, \Delta u \,\mathrm{d}\Gamma = \int_{\Gamma_1} (u - \bar{u}) \frac{\partial \Delta u}{\partial n} \mathrm{d}\Gamma \quad (1.37)$$

Note that from Green's theorem,

$$\int_{\Omega} (u \nabla^2 w - w \nabla^2 u) \,\mathrm{d}\Omega = \int_{\Gamma} \left(\frac{\partial u}{\partial n} w - u \frac{\partial w}{\partial n} \right) \mathrm{d}\Gamma \quad (1.38)$$

where $\Gamma = \Gamma_1 + \Gamma_2$.

With $\nabla^2 w \equiv \nabla^2 u \equiv 0$, and $w = u$ we can write,

$$\int_{\Gamma} q\, \Delta u \,\mathrm{d}\Gamma = \int_{\Gamma} \frac{\partial \Delta u}{\partial n} u \,\mathrm{d}\Gamma \quad (1.39)$$

This relationship can be specialised to obtain equation (1.36).

POISSON'S EQUATION

Note that for the case of Poisson's equation, i.e.

$$\nabla^2 u = b \quad (1.40)$$

the starting weighting residual statement is,

$$\int_{\Omega} (\nabla^2 u - b) w \,\mathrm{d}\Omega = \int_{\Gamma_2} (q - \bar{q}) w \,\mathrm{d}\Gamma - \int_{\Gamma_1} (u - \bar{u}) \frac{\partial w}{\partial n} \mathrm{d}\Gamma \quad (1.41)$$

Integrating by parts,

$$\int_{\Omega} b w \,\mathrm{d}\Omega + \int_{\Gamma_1} u \frac{\partial w}{\partial n} \mathrm{d}\Gamma = \int_{\Gamma} w \frac{\partial u}{\partial n} \mathrm{d}\Gamma \quad (1.42)$$

where $\Gamma = \Gamma_1 + \Gamma_2$. When the same functions are used for w and u, we obtain

$$\int_{\Omega} b\, \Delta u \,\mathrm{d}\Omega + \int_{\Gamma} u \frac{\partial \Delta u}{\partial n} \mathrm{d}\Gamma = \int_{\Gamma} \Delta u \frac{\partial u}{\partial n} \mathrm{d}\Gamma \quad (1.43)$$

This expression contains one domain integral in terms of b but we will see that this term does not introduce any internal unknowns.

Example 1.7

Consider again the same equation as seen in Example 1.6 but now we take,

$$w = \Delta u = \Delta\alpha_1 \cos x + \Delta\alpha_2 \sin x \tag{a}$$

The starting relationship is,

$$\int_0^1 x\,\Delta u\,\mathrm{d}x + [q\,w]_1 - [q\,w]_0 = 0 \tag{b}$$

which gives as previously,

$$q_0 = \frac{1}{\sin 1} - 1, \qquad q_1 = \frac{\cos 1}{\sin 1} - 1 \tag{c}$$

To find the internal values of u in the interval $x = 0$ to 1 we notice that the equation

$$\frac{\mathrm{d}^2 u}{\mathrm{d}x^2} + u = -x \tag{d}$$

may be reduced to homogeneous form

$$\frac{\mathrm{d}^2 v}{\mathrm{d}x^2} + v = 0 \tag{e}$$

by the substitution

$$v = u + x \tag{f}$$

Then assuming an expansion for v of the same form as that for w, i.e.

$$v = \alpha_1 \cos x + \alpha_2 \sin x \tag{g}$$

we can apply the boundary conditions $v = 0$ at $x = 0$, $v = 1$ at $x = 1$ to obtain

$$\alpha_1 = 0 \qquad \alpha_2 = \frac{1}{\sin 1} \tag{h}$$

This gives the correct values for q_0 and q_1 and the exact solution for u, i.e.

$$u = \frac{\sin x}{\sin 1} - x \tag{i}$$

To find values of u at an internal point x_i it is usual, in boundary elements, to choose the fundamental solution for our weighting function (See Section 2.2 and Chapter 4). This is the solution to the equation

$$\frac{\mathrm{d}^2 w^*}{\mathrm{d}x^2} + w^* = \delta_i \tag{j}$$

which is given by

$$w^* = \tfrac{1}{2}\sin r \tag{k}$$

where

$$r = |x - x_i| \tag{l}$$

If we wish to find u at x_i we start from the weighted residual expression

$$\int_0^1 \left\{\frac{\mathrm{d}^2 w^*}{\mathrm{d}x^2} + w^*\right\} u\,\mathrm{d}x + \int_0^1 x\,w^*\,\mathrm{d}x + \Big[q\,w^*\Big]_{x=0}^{x=1} = 0 \tag{m}$$

(equation (c), Example 1.6) which gives by the selective property of the delta function the equation

$$u_i = -\int_0^1 x\,w^*\,\mathrm{d}x - \Big[q\,w^*\Big]_{x=0}^{x=1} \tag{n}$$

If we choose x_i to be at the mid-point of our interval we obtain the value of u at $x = \frac{1}{2}$:

$$u(\tfrac{1}{2}) = -\tfrac{1}{2}\int_0^{1/2} x\sin(\tfrac{1}{2} - x)\,\mathrm{d}x - \tfrac{1}{2}\int_{1/2}^{1} x\sin(x - \tfrac{1}{2})\,\mathrm{d}x - q_1 \sin\tfrac{1}{2} + $$
$$+ q_0 \sin\tfrac{1}{2} \tag{o}$$

As

$$r = \begin{cases} \frac{1}{2} - x & \text{for } 0 < x < \frac{1}{2} \\ x - \frac{1}{2} & \text{for } \frac{1}{2} < x < 1 \end{cases}$$

we obtain

$$u(\tfrac{1}{2}) = -\tfrac{1}{2}[(\cos 1 - 1)/\sin 1] \sin\tfrac{1}{2} - \tfrac{1}{2}(\tfrac{1}{2} - \sin\tfrac{1}{2}) - \tfrac{1}{2}(\tfrac{1}{2} + \sin\tfrac{1}{2} - \cos\tfrac{1}{2})$$
$$= 0{\cdot}069746964 \tag{p}$$

which is the same as the exact solution, i.e.

$$u(\tfrac{1}{2}) = \frac{\sin\frac{1}{2}}{\sin 1} - \tfrac{1}{2} = 0{\cdot}069746964$$

Example 1.8

Consider the following Poisson's equation:

$$\nabla^2 u - b = 0 \tag{a}$$

with homogeneous boundary conditions $u = 0$ at $x = \pm 1$ and $y = \pm 1$.

We can first reduce equation (a) to a Laplace's equation by defining a new function v such that,

$$u = \tfrac{1}{4}b(x^2 + y^2) + v \tag{b}$$

Substituting (b) into (a) we find that,

$$\nabla^2 u - b = \nabla^2 v = 0 \tag{c}$$

with $v = -\frac{1}{4}b(x^2 + y^2)$ on the boundaries $x = \pm 1$ and $y = \pm 1$.

We can now approximate v by a trial function that satisfies the Laplace's equation. For instance,

$$v = (x^4 - 6x^2y^2 + y^4) \tag{d}$$

Because of symmetry we only need to consider a part of the boundary, say $x = 1, 0 < y < 1$. Noting that all the boundary is of Γ_1 type, equation (7) becomes,

$$\int_\Gamma \bar{v} \frac{\partial \Delta v}{\partial n} \, d\Gamma = \int_\Gamma \frac{\partial v}{\partial n} \Delta v \, d\Gamma \tag{e}$$

Taking into consideration symmetry this reduces to,

$$\int_0^1 \left(\bar{v} \frac{\partial \Delta v}{\partial n} - \frac{\partial v}{\partial n} \Delta v \right) \mathrm{d}\Gamma = 0 \tag{f}$$

Substituting (d) into (f) we obtain the following equation:

$$\int_0^1 [\tfrac{1}{4}b(1+y^2)(4-12y^2)\Delta\alpha + \alpha(4-12y^2)(1-6y^2+y^4)\Delta\alpha]\,\mathrm{d}y = 0 \tag{g}$$

Hence,

$$\int_0^1 [\tfrac{1}{4}b(1+y^2) + \alpha(1-6y^2+y^4)](4-12y^2)\,\mathrm{d}y = 0 \tag{h}$$

This gives after integration,

$$-\tfrac{1}{5}b + \alpha\tfrac{144}{35} = 0 \tag{i}$$

$$\therefore \alpha = \tfrac{7}{144}b \tag{j}$$

Whence the value of v becomes,

$$v = \alpha(x^4 - 6x^2y^2 + y^4) = \tfrac{7}{144}b\,(x^4 - 6x^2y^2 + y^4) \tag{k}$$

and u is given by formula (b).

Note that the magnitude of u or v may be displaced from the correct solution by a constant value. In order to determine this value one can carry out the following integration over the $x = 1$, $0 < y < 1$ side:

$$\int_0^1 (\bar{v} - v - c)\,\mathrm{d}y = 0 \tag{l}$$

or,

$$\int_0^1 [-\tfrac{1}{4}b(1+y^2) + \alpha(1-6y^2+y^4) - c]\,\mathrm{d}y = 0 \tag{m}$$

Thus,

$$c = -\tfrac{165}{432}b \tag{n}$$

Example 1.9

Let us now solve the same problem as in Example 1.8, but using boundary collocation and starting with equation (9). We can take as an approximate solution the same function as in the previous example, i.e.

$$u = \tfrac{1}{4}b(x^2+y^2)+\alpha_1+\alpha_2(x^4-6x^2y^2+y^4) \qquad \text{(a)}$$

and collocate it over the boundary $x = 1$, $0 < y < 1$. The starting expression is

$$\int_{\Gamma_1} (u-\bar{u})\frac{\partial \Delta u}{\partial n}\,\mathrm{d}\Gamma = 0 \qquad \text{(b)}$$

or

$$\int_0^1 (u-\bar{u})\frac{\partial \Delta u}{\partial n}\,\mathrm{d}\Gamma = 0 \qquad \text{(c)}$$

which for two collocation points on the boundary reduces to,

$$\int_0^1 (u-\bar{u})\,\delta_i\,\mathrm{d}\Gamma = 0 \qquad i = 1, 2 \qquad \text{(d)}$$

Let us take the two points on $x = 1$, one at $y = 0{\cdot}25$ and the other at $y = 0{\cdot}75$. This will give the following system of equations:

$$\begin{bmatrix} 1 & 0{\cdot}664 \\ 1 & -2{\cdot}058 \end{bmatrix}\begin{Bmatrix} \alpha_1 \\ \alpha_2 \end{Bmatrix} = -\begin{Bmatrix} 0{\cdot}2656 \\ 0{\cdot}3906 \end{Bmatrix} b \qquad \text{(e)}$$

with,

$$\alpha_1 = -0{\cdot}2961\,b, \qquad \alpha_2 = 0{\cdot}0459\,b \qquad \text{(f)}$$

This gives

$$u = \tfrac{1}{4}b(x^2+y^2)-0{\cdot}2961b+0{\cdot}0459b(x^4-6x^2y^2+y^4) \qquad \text{(g)}$$

Equation (g) gives a u value different from the one obtained in Example 1.8. For instance, if we compare the values of u at the centre ($x = y = 0$) for the two examples we have,

$$\begin{aligned} &\text{previous example} \quad u_c = -0{\cdot}3819b \\ &\text{this example} \qquad\quad u_c = -0{\cdot}2961b \end{aligned} \qquad \text{(h)}$$

REFERENCES

1. Lau, P. and C. A. Brebbia, 'The Cell Collocation Method', *Int. J. Mech. Sci.*, **20,** 83–95 (1978)

BIBLIOGRAPHY

Brebbia, C. A. and Dominguez, J., 'The Boundary Element Method for Potential Problems', *Appl. Math. Modelling,* **1,** No. 7, December (1977).

Brebbia, C. A. and Connor, J. J., *Finite Elements for Fluid Flow*, Butterworths, second edition (1977)

Brebbia, C. A. and Connor, J. J., *Fundamentals of Finite Elements for Structural Engineers*, Butterworths (1973)

2 Potential Problems

2.1 INTRODUCTION

Many engineering problems such as seepage, heat conduction, electrical problems, etc. are governed by a Laplace or Poisson equation of the type,

$$\nabla^2 u = b \tag{2.1}$$

with b some known function and with boundary conditions (see Figure 2.1)

$$\begin{aligned} u &= \bar{u} && \text{on } \Gamma_1 \\ q &= \partial u/\partial n = \bar{q} && \text{on } \Gamma_2 \end{aligned} \tag{2.2}$$

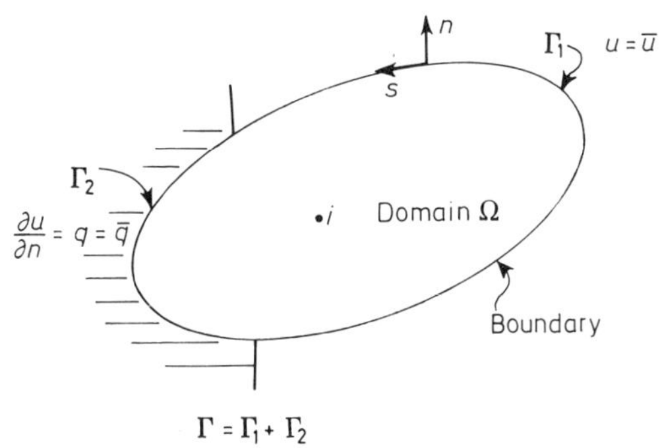

Figure 2.1 Notation for potential problems

As we have seen, this system can be written as a weighted residual statement as follows,

$$-\int_{\Omega} bw\,d\Omega + \int_{\Omega} (\nabla^2 u)w\,d\Omega = \int_{\Gamma_2} (q-\bar{q})w\,d\Gamma - \int_{\Gamma_1} (u-\bar{u})\frac{\partial w}{\partial n}\,d\Gamma \tag{2.3}$$

Integrating this expression by parts twice we obtain,

$$-\int_{\Omega} bw\,d\Omega + \int_{\Omega} (\nabla^2 w)u\,d\Omega = -\int_{\Gamma_2} \bar{q}\,w\,d\Gamma - \int_{\Gamma_1} qw\,d\Gamma + \int_{\Gamma_2} u\frac{\partial w}{\partial n}\,d\Gamma + \int_{\Gamma_1} \bar{u}\frac{\partial w}{\partial n}\,d\Gamma \tag{2.4}$$

Note that in order to transform the problem into a boundary problem we need to find a w function such that $\nabla^2 w = 0$ or a u function giving $\nabla^2 u = 0$. These approaches leave us with only one domain integral in terms of b. We can divide the boundary solutions into two types,

(a) Solutions which satisfy the equation $\nabla^2 u = 0$ in Ω but not the boundary conditions on Γ. These solutions can be found with equation (2.3).
(b) Methods for which the weighting function satisfies $\nabla^2 w = 0$. The boundary terms need weighting as shown in equation (2.4).

The two methods of solution may be based on approximate functions that satisfy $\nabla^2(\quad) = 0$ or on a particular type of solution called the fundamental solution. This is the solution for the Laplace equation in an infinite domain and with a unit applied potential at a given point 'i', i.e.

$$\nabla^2 u^* = \delta_i \tag{2.5}$$

where δ_i is a Dirac delta function representing a unit concentrated potential acting at a point i. This type of solution is widely used in boundary problems and represents the Green's or influence function.

Solutions that satisfy $\nabla^2 u \equiv 0$ have been shown in Chapter 1. We are now interested in the possibility of using the fundamental solution to develop a more general method.

2.2 THE FUNDAMENTAL SOLUTION AND DIRECT FORMULATION

The solution corresponding to an applied potential concentrated at a point is frequently used in boundary problems. For the Laplace's equation for instance the fundamental solution that will be called u^* (the asterisk is used to indicate its special character) is the one corresponding to the following equation:

$$\nabla^2 u^* = \delta_i \tag{2.6}$$

or more explicitly

$$\nabla^2 u^* = \delta(\vec{x} - \vec{x}_i) \tag{2.7}$$

where δ_i is the Dirac delta function.

Note that this function u^* is a function of two points: the 'source' point $\vec{x}_i$ at which we have the singularity of our delta function; and the 'observation' point $\vec{x}$ which is the variable involved in our differential equation.

For equations such as (2.6) or (2.7) the fundamental solution is a function only of the distance between the source point and the observation point. We denote this distance by r defined by (see Figure 2.2)

$$r = |\vec{r}| = |\vec{x} - \vec{x}_i| \tag{2.8}$$

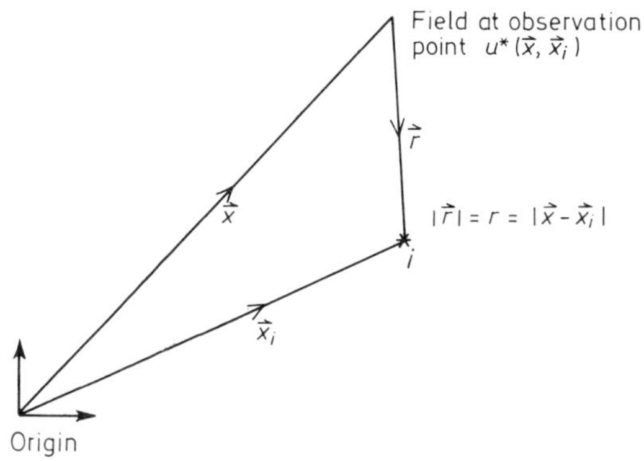

Figure 2.2 Notation for definition of the fundamental solution

This function when weighted by any other function u, has the following property:

$$\int_{\Omega} (\nabla^2 u^*) u \, \mathrm{d}\Omega = \int_{\Omega} \delta_i u \, \mathrm{d}\Omega = u_i \tag{2.9}$$

u_i represents the value of the unknown function u at the point of application of the potential.

The fundamental solution of equation (2.6) for a three-dimensional isotropic medium is,

$$\nabla^2 u^* = 1/4\pi r \tag{2.10}$$

We can show that (2.10) is the fundamental solution required by considering equation (2.6) in spherical polar coordinates.

The field due to a point source in isotropic space will only be a function of the radial coordinate r, taken to be the distance from the source, hence we can write

$$\frac{\partial^2 u^*}{\partial r^2} + \frac{2}{r} \frac{\partial u^*}{\partial r} + \delta_0 = 0 \tag{2.11}$$

without loss of generality we have now taken the source point to be at the origin. Substituting (2.10) into (2.11) we can easily verify that (2.10) satisfies the governing equation for any r different from zero. At $r = 0$ a singularity occurs and we have to consider a small sphere surrounding the point i where the load is applied and integrate over it to find out what will occur when $r \to 0$.

This can be done by using Gauss' theorem in the form

$$\int_{\Omega_e} \nabla^2 u^* \, \mathrm{d}\Omega = \int_{\Gamma_e} \frac{\partial u^*}{\partial r} \mathrm{d}\Gamma \tag{2.12}$$

where Ω_e, Γ_e are the volume and surface of the sphere surrounding the source point. Substituting (2.10) into (2.12) we obtain,

$$\int_{\Gamma_e} \frac{\partial u^*}{\partial r} \mathrm{d}\Gamma = \frac{1}{4\pi} \int_{\Gamma_e} \left(-\frac{1}{r^2} \right) \mathrm{d}\Gamma_e = -\frac{1}{4\pi} \left(\frac{4\pi r^2}{r^2} \right)_{\Gamma_e} = -1 \tag{2.13}$$

Note that this result is independent of r, hence when $r \to 0$, the integral of the potential over the sphere tends to 1. We have therefore represented a unit source strength.

The fundamental solution for the isotropic two-dimensional case is

$$u^* = \frac{1}{2\pi} \ln\left(\frac{1}{r}\right) \tag{2.14}$$

and has similar properties to the three-dimensional case.

If we assume that the weighting function is composed of a series of fundamental solutions applied over the boundary we can work with equation (2.4). Conversely if the approximating function is assumed to be a combination of fundamental solutions, equation (2.3) could be used. (Both functions can also be taken to be combinations of fundamental solutions as in the case $w = \Delta u$.)

For the first case we can consider a particular point 'i' for which u^* is the fundamental solution due to an applied potential acting at that point. This equation can be written as,

$$-\int_{\Omega} b u^* \,\mathrm{d}\Omega + \int_{\Omega} (\nabla^2 u^*) u \,\mathrm{d}\Omega + \int_{\Gamma_2} \bar{q} u^* \,\mathrm{d}\Gamma + \int_{\Gamma_1} q u^* \,\mathrm{d}\Gamma = \int_{\Gamma_1} \bar{u} q^* \,\mathrm{d}\Gamma + \int_{\Gamma_2} u q^* \,\mathrm{d}\Gamma \tag{2.15}$$

where $q^* = \partial u^* / \partial n$.

Taking into consideration that at 'i' the governing equation for u^* is

$$\nabla^2 u^* = \delta_i \tag{2.16}$$

Formula (2.15) for the point 'i' under consideration becomes,

$$u_i + \int_{\Omega} b u^* \,\mathrm{d}\Omega + \int_{\Gamma_2} u q^* \,\mathrm{d}\Gamma + \int_{\Gamma_1} \bar{u} q^* \,\mathrm{d}\Gamma = \int_{\Gamma_2} \bar{q} u^* \,\mathrm{d}\Gamma + \int_{\Gamma_1} q u^* \,\mathrm{d}\Gamma \tag{2.17}$$

This equation relates the value of u at the point 'i' with the values of q and u over the boundary Γ (Figure 2.1). It appplies when the point 'i' is in the interior domain Ω, but in order to formulate the problem in terms of boundary integrals we need to take the point 'i' to the boundary. We will now do this analytically for a smooth boundary though we will later on see that we do not need to know the analytical expression.

Consider a hemisphere on the boundary of a three-dimensional domain as shown in Figure 2.3. (The two-dimensional case can be analysed in a similar way.) The boundary point can be assumed to be at the centre of the sphere and afterwards the radius 'ε' is reduced to zero. The point will then become a boundary point.

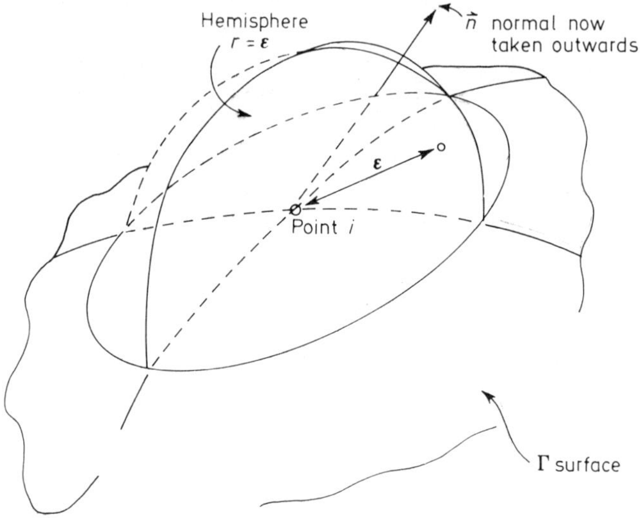

Figure 2.3 Boundary point surrounded by a hemisphere

Let us assume that the boundary is of the type Γ_2 and consider it to be divided into two parts, i.e.

$$\int_{\Gamma_2} u\, q^*\, \mathrm{d}\Gamma = \int_{\Gamma_{(2-\varepsilon)}} u\, q^*\, \mathrm{d}\Gamma + \int_{\Gamma_\varepsilon} u\, q^*\, \mathrm{d}\Gamma \tag{2.18}$$

Consider the second term and replace q^* by its value in terms of the fundamental solution. This gives,

$$\int_{\Gamma_\varepsilon} u\, q^*\, \mathrm{d}\Gamma = -\int_{\Gamma_\varepsilon} u \frac{1}{4\pi\varepsilon^2}\, \mathrm{d}\Gamma \tag{2.19}$$

Knowing that the area of the hemisphere is equal to $2\pi\varepsilon^2$ we can take

(2.19) to the limit, i.e. let $\varepsilon \to 0$ and $u \to u_i$, which gives

$$\lim_{\varepsilon \to 0} \left(-\int_{\Gamma_\varepsilon} u \frac{1}{4\pi\varepsilon^2} \mathrm{d}\Gamma \right) = \lim_{\varepsilon \to 0} \left(-\tfrac{1}{2}u_i \right) = -\tfrac{1}{2}u_i \tag{2.20}$$

Note that as the ε is now zero the boundary $\Gamma_{(2-\varepsilon)}$ again can be represented as Γ_2 and that we have integrated over the singularity, which gives $-\frac{1}{2}u_i$. The same result $(-\frac{1}{2}u_i)$ is obtained in two-dimensional problems with smooth boundaries.

The right-hand side integrals in equation (2.17) do not give any singularity. This can be shown by taking the limit of the following integral when $\varepsilon \to 0$, i.e.

$$\lim_{\varepsilon \to 0} \left(\int_{\Gamma_\varepsilon} q\, u^* \,\mathrm{d}\Gamma \right) = \lim_{\varepsilon \to 0} \left(\int_{\Gamma_\varepsilon} q \frac{1}{4\pi\varepsilon} \mathrm{d}\Gamma \right) = 0 \tag{2.21}$$

Substituting the result given by equation (2.20) into our original relationship (2.17) we obtain the following for a point on the boundary:

$$\tfrac{1}{2}u_i + \int_\Omega b u^* \,\mathrm{d}\Omega + \int_{\Gamma_2} uq^* \,\mathrm{d}\Gamma + \int_{\Gamma_1} \bar{u}q^* \,\mathrm{d}\Gamma = \int_{\Gamma_2} \bar{q}u^* \,\mathrm{d}\Gamma + \int_{\Gamma_1} qu^* \,\mathrm{d}\Gamma \tag{2.22}$$

The same result would be obtained if instead of considering the point on the Γ_2 part of the boundary, we considered it to be on the Γ_1 part.

Equation (2.22) can be written in a more compact way as follows,

$$\tfrac{1}{2}u_i + \int_\Omega b u^* \,\mathrm{d}\Omega + \int_\Gamma uq^* \,\mathrm{d}\Gamma = \int_\Gamma qu^* \,\mathrm{d}\Gamma \tag{2.23}$$

where $\Gamma = \Gamma_1 + \Gamma_2$ and we then apply the boundary conditions $u = \vec{u}$ on Γ_1 and $q = \vec{q}$ *on* Γ_2.

2.3 THE INDIRECT FORMULATION

It is frequent in boundary solutions to use a source or dipole distribution on the boundary and solve the problem in terms of unknown sources or dipoles. This technique is sometimes called the indirect formulation and can be deduced from the more general direct formulation statement (2.17).

Let us assume for a moment that all the boundary is of the Γ_1 type. Hence equation (2.17) becomes,

$$u_i + \int_{\Gamma_1} \bar{u}\, q^*\, \mathrm{d}\Gamma = \int_{\Gamma_1} q\, u^*\, \mathrm{d}\Gamma \tag{2.24}$$

Brebbia and Butterfield[1] demonstrated how the direct statement can be transformed into an equivalent indirect statement for the case of Poisson's equation and general self-adjoint operators.

Let us consider the Poisson's equation:

$$\nabla^2 u + b = 0 \tag{2.25}$$

with the fundamental solution u^*, which is the solution of

$$\nabla^2 u^* + \delta_i = 0 \tag{2.26}$$

The basic relationship for the direct method is,

$$c_i u_i + \int_{\Gamma} u q^*\, \mathrm{d}\Gamma = \int_{\Gamma} q u^*\, \mathrm{d}\Gamma - \int_{\Omega} b u^*\, \mathrm{d}\Omega \tag{2.27}$$

where c_i is 1 at an internal point, $\frac{1}{2}$ for a 'smooth boundary' point or 0 for a point outside the region Ω. Let us call this external region Ω' and assume that u' is the solution to the Laplace equation $\nabla^2 u' = 0$ within Ω'.

For this case we obtain instead of equation (2.27) the following statement:

$$\int_{\Gamma} u^* q'\, \mathrm{d}\Gamma + \int_{\Gamma} q^* u'\, \mathrm{d}\Gamma = 0 \tag{2.28}$$

Note that the point 'i' (*observation point*) is now external to Ω and the sign q^* has been changed in equation (2.28) so that the two values of q^* in (2.27) and (2.28) become identical despite the reversal of the unit normal.

We can now specify u' as the solution in Ω' which generates potentials around Γ identical to those of our initial problem, within Ω, that is $u = u'$ on Γ. Hence we can add equations (2.27) and (2.28) which gives, for an internal point in Ω,

$$u_i - \int_\Gamma u^* (q + q')\mathrm{d}\Gamma + \int_\Omega b u^* \,\mathrm{d}\Omega = 0 \tag{2.29}$$

or

$$u_i = \int_\Gamma u^* \sigma \mathrm{d}\Gamma - \int_\Omega b u^* \mathrm{d}\Omega \tag{2.30}$$

This is the form of the indirect boundary element method statement. σ represents the initially unknown density distribution of u^* over Γ necessary to generate u_i via equation (2.30). Notice that in the indirect formulation the singular source u^* is always considered to be located at the boundary point rather than at any internal boundary point or even an external point as in the direct formulation. The physical interpretation of σ is the difference between the fluxes generated by the two solutions (internal and external).

We can also deduce another indirect formulation from equations (2.27) and (2.28). This is the *indirect* approach in terms of dipoles. Instead of continuity of potentials we can assume continuity of fluxes, i.e. $q + q = 0$. Subtracting equations (2.27) and (2.28) we now obtain

$$u_i = -\int_\Gamma (u - u')q^* \mathrm{d}\Gamma - \int_\Omega b u^* \mathrm{d}\Omega \tag{2.31}$$

or

$$u_i = \int_\Gamma \mu q^* \mathrm{d}\Gamma - \int_\Omega b u^* \mathrm{d}\Omega \tag{2.32}$$

μ are called dipoles and give rise to an alternative indirect formulation.

The steps for any self-adjoint case are similar to the ones described for the above equations and they have been described in detail in Brebbia and Butterfield.[1] Similar arguments can also be applied to problems such as those present in elasticity (Chapter 4) and the argument can even be extended to non-self-adjoint problems but for these cases the fundamental solution for the adjoint operator may be difficult to find.

Note that in order to apply natural boundary conditions on Γ, in the source or dipole formulation we need to find the derivatives of (2.27) and (2.30). For the source formulation for instance the flux at any point 'i' is given by,

$$q_i = \left(\frac{\partial u}{\partial n}\right)_i = \int_\Gamma \sigma q^* \mathrm{d}\Gamma - \int_\Omega b q^* \mathrm{d}\Omega \tag{2.33}$$

Note that if the point i is on the boundary we have a singularity in equation (2.30) and we obtain,

$$q_i = -\tfrac{1}{2}\sigma_i + \int_\Gamma \sigma q^* \mathrm{d}\Gamma - \int_\Omega b q^* \mathrm{d}\Omega \tag{2.34}$$

Equations (2.28) and (2.32) are the starting relationships for the source approach. They can be used even if the boundary values are defined in terms of $\bar{u}$ and $\bar{q}$, simply by using (2.27) on the part of the boundary where $\bar{u}$ is prescribed, and (2.34) on the other part. This gives,

$$u_j = \int_\Gamma \sigma u^* \mathrm{d}\Gamma + \int_\Omega p u^* \mathrm{d}\Omega = \bar{u}_j \tag{2.35}$$

$$q_j = -\tfrac{1}{2}\sigma_j + \int_\Gamma \sigma q^* \mathrm{d}\Gamma + \int_\Omega p q^* \mathrm{d}\Omega = \bar{q}_j \tag{2.36}$$

Equations (2.35) and (2.36) are frequently used. Conversely we can use a *dipole* type formulation, but this is slightly more cumbersome as it involves second derivatives of the potential and in consequence is applied less frequently.

Example 2.1

Let us illustrate the source formulation by considering the same second-order equation as previously, i.e.

$$\frac{\mathrm{d}^2 u}{\mathrm{d}x^2} + u + x = 0 \tag{a}$$

The fundamental solution for

$$\frac{\mathrm{d}^2 u^*}{\mathrm{d}x^2} + u^* + \delta_i = 0 \tag{b}$$

is given by

$$u^* = -\tfrac{1}{2}\sin r \tag{c}$$

The source formulation gives

$$u = \int_\Gamma \sigma u^*\, d\Gamma - \int_\Omega b u^*\, d\Omega \tag{d}$$

Equation (d) can be applied at points $x = 0$ and $x = 1$, to obtain,

$$\begin{aligned} u_0 &= \sigma_1 (\tfrac{1}{2}\sin 1) + \tfrac{1}{2}\int_0^1 x \sin x \, dx = 0 \\ u_1 &= \sigma_0 (\tfrac{1}{2}\sin 1) + \tfrac{1}{2}\int_0^1 x \sin (1-x)\, dx = 0 \end{aligned} \tag{e}$$

This system gives,

$$\sigma_0 = \frac{1}{\sin 1} - 1, \qquad \sigma_1 = \frac{\cos 1}{\sin 1} - 1 \tag{f}$$

Note that these two values for the sources are the same as the q_0 and q_1 of Example 1.7.

2.4 MATRIX FORMULATION

We will now explain how the boundary element techniques described in Sections 2.2 and 2.3 can be interpreted in matrix form. The direct case given by relationship (2.23) will be considered in detail. This case is more general than equations (2.35) and (2.36) which can also be rendered into matrix form.

Consider equation (2.23) specialised for the boundary points and without the term in p for simplicity (the case with p is considered later), i.e.

$$\tfrac{1}{2}u_i + \int_\Gamma u q^*\, d\Gamma = \int_\Gamma q u^*\, d\Gamma \tag{2.41}$$

Assume for simplicity that the body is two dimensional and its boundary is divided into N 'segments' or 'boundary' elements, as shown in Figure 2.4. The points where the unknown values are

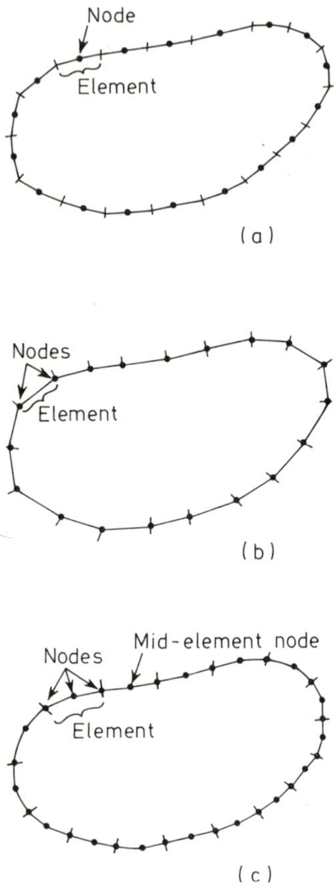

Figure 2.4 Boundary elements: (a) *constant;* (b) *linear;* (c) *quadratic*

considered are called 'nodes' and taken to be in the middle of each segment for the so-called 'constant' elements (Figure 2.4(a)). These are going to be the only elements considered in this section, but in the next chapter we will also discuss the case of *linear* elements (i.e. elements for which the nodes are at the intersection between elements as shown in Figure 2.4(b)) and curved elements such as the quadratic ones shown in Figure 2.4(c). For the latter case an extra mid-element node is required and the elements are called quadratic.

For the constant element case the boundary is discretised into N elements, of which N_1 are assumed to belong to Γ_1 and N_2 to Γ_2. The values of u and q are assumed to be constant on each element and equal to the values at the mid-node of the element. Note that on each element one of the two variables (u or q) is known.

Equation (2.41) can be discretised for a given point 'i' before applying any boundary conditions, as follows,

$$\tfrac{1}{2}u_i + \sum_{j=1}^{N} \int_{\Gamma_j} u\, q^*\, \mathrm{d}\Gamma = \sum_{j=1}^{N} \int_{\Gamma_j} q\, u^*\, \mathrm{d}\Gamma \qquad (2.42)$$

Note that for this type of element the boundary is always 'smooth' hence the multiplier of u_i is $\tfrac{1}{2}$. Γ_j is the boundary of element 'j'.

Formula (2.42) applies for a particular node 'i'. The u_j and q_j values can be taken out of the integrals as they are assumed to be constant over each element. This gives,

$$\tfrac{1}{2}u_i + \sum_{j=1}^{N} \left(\int_{\Gamma_j} q^*\, \mathrm{d}\Gamma \right) u_j = \sum_{j=1}^{N} \left(\int_{\Gamma_j} u^*\, \mathrm{d}\Gamma \right) q_j \qquad (2.43)$$

The integrals $\int q^*\, \mathrm{d}\Gamma$ relate the 'i' node with the segment 'j' over which the integral is carried out. We shall call these integrals $\hat{H}_{ij}$. The integrals on the right-hand side are of the type $\int u^*\, \mathrm{d}\Gamma$ and will be called G_{ij}. Hence we can write (2.43) as,

$$\tfrac{1}{2}u_i + \sum_{j=1}^{N} \hat{H}_{ij} u_j = \sum_{j=1}^{N} G_{ij} q_j \qquad (2.44)$$

The integrals can easily be calculated analytically but for higher-order elements they become more difficult to evaluate and they are usually evaluated numerically.

We can rewrite equation (2.44) for each i node under consideration. Let us now call

$$H_{ij} = \begin{cases} \hat{H}_{ij} & \text{when } i \neq j \\ \hat{H}_{ij} + \tfrac{1}{2} & \text{when } i = j \end{cases} \qquad (2.45)$$

Hence equation (2.44) can now be written as,

$$\sum_{j=1}^{N} H_{ij} u_j = \sum_{j=1}^{N} G_{ij} q_j \qquad (2.46)$$

The whole set of equations can also be expressed in matrix form as,

$$\mathbf{H}\mathbf{U} = \mathbf{G}\mathbf{Q} \qquad (2.47)$$

Note that N_1 values of u and N_2 values of q are known on Γ, hence one has a set of only N unknows in formula (2.47) which can be reordered according with the unknown under consideration. We can pass all unknowns to the left-hand side and write,

$$\mathbf{AX} = \mathbf{F} \tag{2.48}$$

where $\mathbf{X}$ is the vector of unknown u's and q's.

Equation (2.48) can now be solved and we then know all the u's and q's values on the boundary. Once this is done we can calculate the values of u's and q's at any interior point using equation (2.17), i.e.

$$u_i = \int_\Gamma q\, u^* \,\mathrm{d}\Gamma - \int_\Gamma u\, q^* \,\mathrm{d}\Gamma \tag{2.49}$$

The u's can be obtained directly from (2.49) by discretising it as follows,

$$u_i = \sum_{j=1}^{N} q_j G_{ij} - \sum_{j=1}^{N} u_j \hat{H}_{ij} \tag{2.50}$$

The values of internal fluxes can be calculated by differentiating (2.49), which gives (see Figure 2.5) for a point i

$$\begin{aligned} q_x &= \frac{\partial u}{\partial x} = \int_\Gamma q \frac{\partial u^*}{\partial x} \mathrm{d}\Gamma - \int_\Gamma u \frac{\partial q^*}{\partial x} \mathrm{d}\Gamma \\ q_y &= \frac{\partial u}{\partial y} = \int_\Gamma q \frac{\partial u^*}{\partial y} \mathrm{d}\Gamma - \int_\Gamma u \frac{\partial q^*}{\partial y} \mathrm{d}\Gamma \end{aligned} \tag{2.51}$$

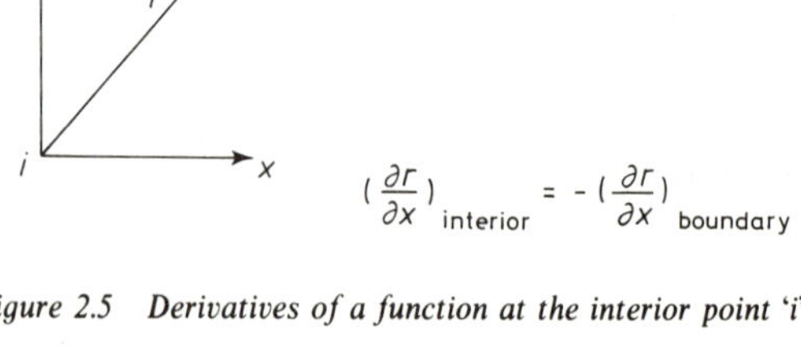

Figure 2.5 Derivatives of a function at the interior point 'i'

EVALUATION OF INTEGRALS

The integrals $\hat{H}_{ij}$ and G_{ij} can be calculated using the simple Gauss quadrature rule (four points for the two-dimensional case) for all points, except the one corresponding to the node under consideration. For this integral it is recommended to use higher-order integration rules or a special logarithmic formula which will be described later on. For the particular case of constant elements, however, the $\hat{H}_{ii}$ and G_{ii} integrals can be easily computed analytically. The $\hat{H}_{ii}$ term for instance is identically zero, for fundamental solutions with no Γ dependence, i.e.

$$\hat{H}_{ii} = \int_{\Gamma_i} q^* \, d\Gamma = \int_{\Gamma_i} \frac{\partial u^*}{\partial r} \frac{\partial r}{\partial n} d\Gamma \equiv 0 \tag{2.52}$$

This is because n and r are orthogonal over the element.

The G_{ii} integral can be calculated analytically as follows,

$$G_{ii} = \int_{\Gamma_{ii}} u^* \, d\Gamma = \frac{1}{2\pi} \int_{\Gamma_i} \ln\left(\frac{1}{r}\right) d\Gamma \tag{2.53}$$

or using the homogeneous coordinate ξ over a segment (Figure 2.6):

$$G_{ii} = \frac{1}{2\pi} \int_{(1)}^{(2)} \ln\left(\frac{1}{r}\right) d\Gamma = \frac{1}{\pi} \int_{(0)}^{(2)} \ln\left(\frac{1}{r}\right) d\Gamma \tag{2.54}$$

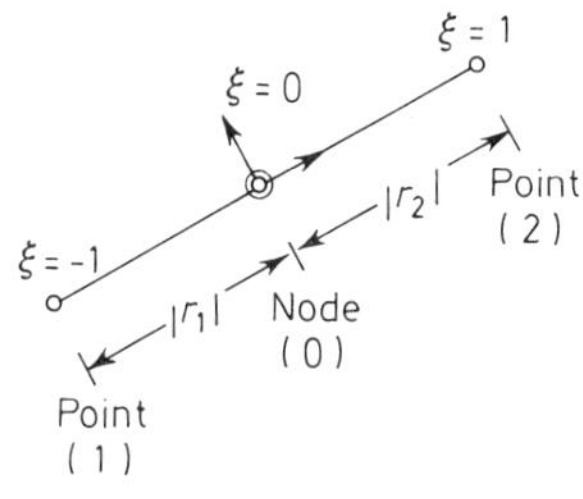

Figure 2.6 Constant element

Transforming coordinates, $r = \xi |r_1|$, where $|r_1| = |r_2|$ we have,

$$G_{ii} = \frac{1}{\pi} \int_{(0)}^{(2)} \ln\left(\frac{1}{r}\right) d\Gamma = \frac{|r_1|}{\pi} \left[\ln\left(\frac{1}{|r_1|}\right) + \int_0^1 \ln\left(\frac{1}{\xi}\right) d\xi \right] \tag{2.55}$$

Noting that the last integral is equal to 1 we have that,

$$G_{ii} = \frac{1}{\pi}|r_1|\left[\ln\left(\frac{1}{|r_1|}\right)+1\right] \tag{2.56}$$

Numerical integration is preferred when working with more complex elements and a specially logarithmically weighted integration formula can be used (see Stroud and Secrest[2]) This formula expresses a logarithmic integral as follows,

$$\int_0^1 \ln\left(\frac{1}{r}\right) f(r)\,\mathrm{d}r = \sum_{i=1}^{k} A_i(r_i) \tag{2.57}$$

k is the number of integration points.

THE SOURCE FORMULATION

The source formulation can equally well be expressed in matrix form, starting with equations (2.35) and (2.36), without the b term for simplicity, i.e.

$$\begin{aligned} u_i &= \int_\Gamma \sigma u^* \,\mathrm{d}\Gamma \\ q_k &= -\tfrac{1}{2}\sigma_k + \int_\Gamma \sigma q^* \,\mathrm{d}\Gamma \end{aligned} \tag{2.58}$$

Discretising (2.58) we obtain,

$$\begin{aligned} u_i &= \sum_{j=1}^{N} \sigma_j \int_{\Gamma_j} u^* \,\mathrm{d}\Gamma \\ q_k &= -\tfrac{1}{2}\sigma_k + \sum_{j=1}^{N} \sigma_j \int_{\Gamma_j} q^* \,\mathrm{d}\Gamma \end{aligned} \tag{2.59}$$

Note that the integrals can again be written as $\hat{H}_{ij}$ and G_{ij}, i.e.

$$\begin{aligned} u_i &= \sum_{j=1}^{N} \sigma_j G_{ij} \\ q_k &= -\tfrac{1}{2}\sigma_k + \sum_{j=1}^{N} \sigma_j \hat{H}_{ij} \end{aligned} \tag{2.60}$$

The difference is now that the G and H contributions are uncoupled and that the diagonal terms in H_{ij} are

$$H_{ii} = \hat{H}_{ii} - \tfrac{1}{2} \tag{2.61}$$

instead of plus $\frac{1}{2}$ (see equation (2.45)).

Formula (2.60) can now be written as,

$$\begin{aligned} u_i &= \sum_{j=1}^{N} \sigma_j G_{ij} \\ q_k &= \sum_{j=1}^{N} \sigma_j H_{kj} \end{aligned} \qquad (2.62)$$

Equations (2.62) gives a system of $2N$ equations which can now be reduced to a $N \times N$ system if the necessary boundary conditions are applied,

$$\begin{aligned} u_i &= \bar{u}_i \quad \text{at } N_1 \text{ points on } \Gamma_1 \\ q_k &= \bar{q}_k \quad \text{at } N_2 \text{ points on } \Gamma_2 \end{aligned} \qquad (2.63)$$

The final system can then be written as,

$$\mathbf{A}\boldsymbol{\sigma} = \mathbf{F} \qquad (2.64)$$

where the unknowns in the $\boldsymbol{\sigma}$ vector are the source intensities.

It is important to point out that an 'equilibrium' type requirement should be introduced in two-dimensional problems. The reason is that the potentials given by (2.62) are only relative potentials in two-dimensional problems. This is because of the special logarithmic character of the solution, i.e. $u^* = -(2\pi)^{-1} \ln(1/r)$, which implies that u^* does not go to zero at infinity.

As the choice of datum is arbitrary we can write

$$u_i = \sum_{j=1}^{N} \sigma_j G_{ij} + C \qquad (2.65)$$

where C is an unknown constant. The system has now $(N+1)$ unknowns but we can solve it by adding the following 'equilibrium' condition:

$$\int_{\Gamma} \sigma \, \mathrm{d}\Gamma = 0 \qquad (2.66)$$

or in discrete form,

$$\sum_{j=1}^{N} \sigma_j l_j = 0 \qquad (2.67)$$

where l_j is the length of each element. This equation represents the requirement that the sum of all the sources applied is zero and will ensure that its effect at infinity is zero. Another way of ensuring this is by closing the domain with a series of boundary elements far away from the body.

Example 2.2

Figure 2.7 depicts the first application of finite elements to fluid flow, i.e. a flow around a cylinder between parallel plates, and is due to Martin.[3] u represents the stream function and $\partial u/\partial n$ are the velocities along the boundaries. The values of the stream function and its derivative are shown in Figure 2.7.

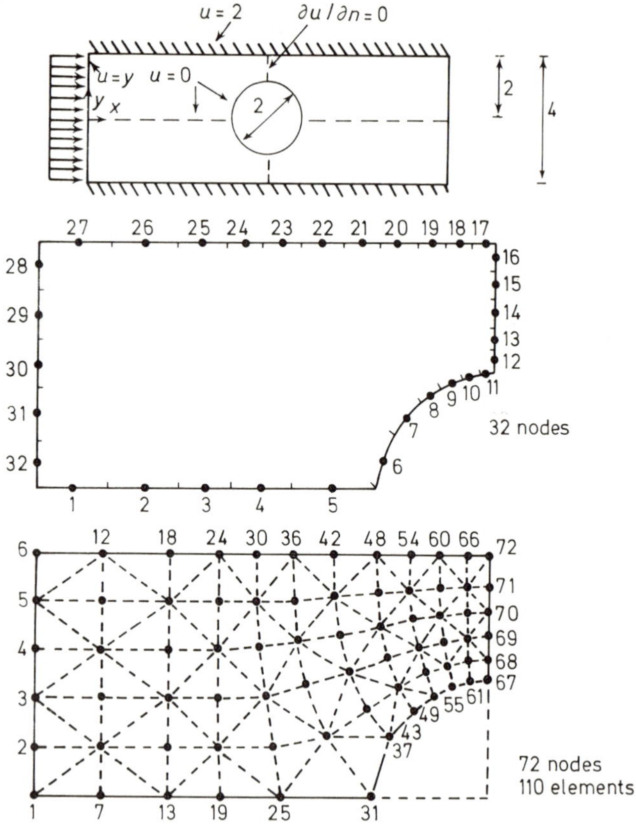

Figure 2.7 Flow around a cylinder between parallel plates. Constant boundary elements versus finite element grid

Because of symmetry we need only to consider a quarter of the domain. This has been discretised using 110 finite elements with 72 nodes. A constant boundary element grid with the same density on the boundaries has only 32 elements or nodes. Figure 2.8(a) shows the prescribed u and $\partial u/\partial n$ values along the boundary and Figure 2.8(b) the computed values for u and $\partial u/\partial n$. Note that the variation of $\partial u/\partial n$ velocities is the one expected.

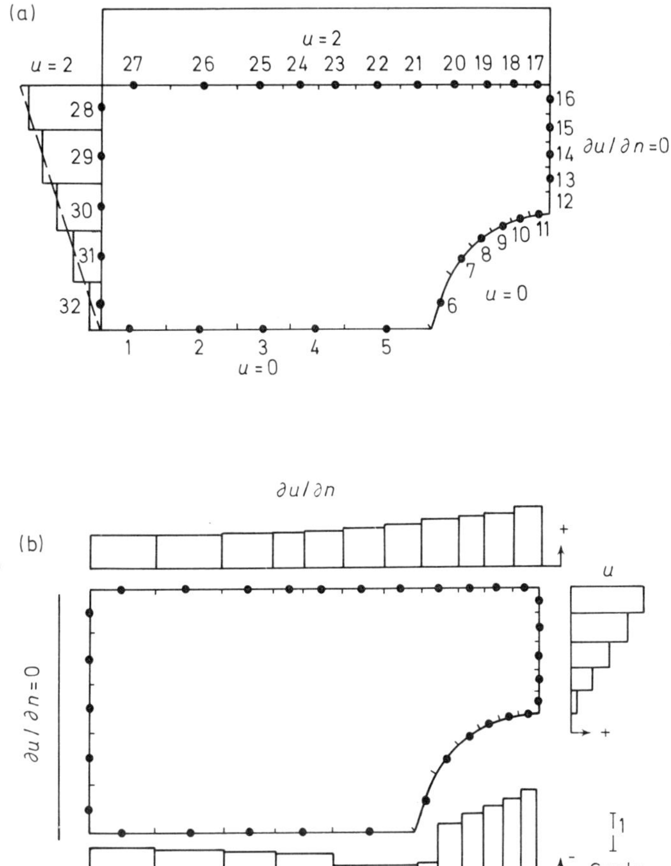

Figure 2.8 Prescribed conditions and solution for constant elements. (a) *Values of u and ∂u/∂n prescribed along boundary;* (b) *computed values for u and ∂u/∂n along boundary*

Values of u inside the domain agree well with the finite elements results and the cell collocation solution (developed in Lau and Brebbia[4]). The cell collocation method for this case is similar to a curvilinear finite difference technique.

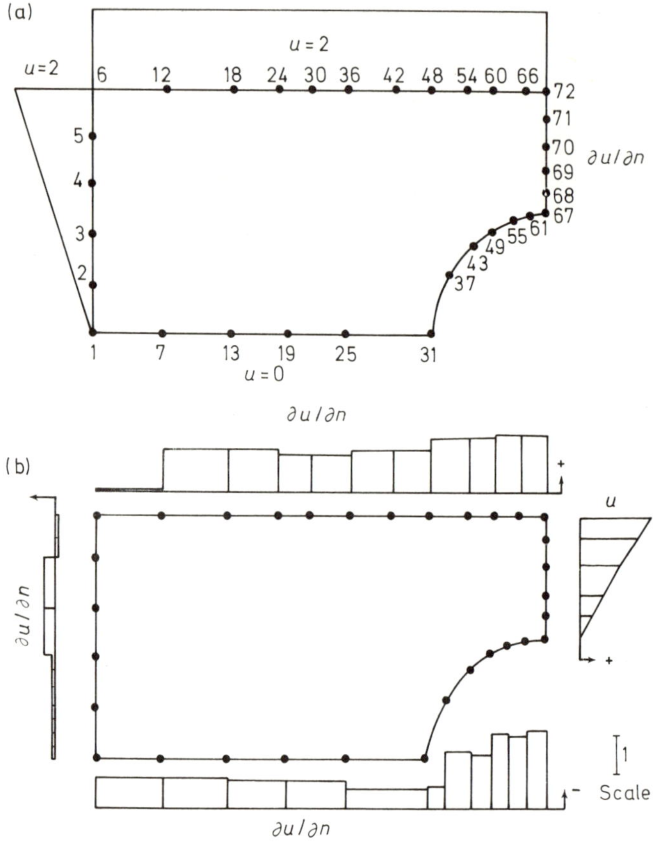

Figure 2.9 Prescribed conditions and solution for finite elements. (a) *Values of u and $\partial u/\partial n$ prescribed along boundary;* (b) *computed values for u and $\partial u/\partial n$ along boundary*

Notice that the values of velocities ($\partial u/\partial n$) calculated on the boundaries using finite elements are unsatisfactory (Figure 2.9), while the boundary element and cell collocation solutions represent the expected flow configuration well. A comparison of values of u for some internal points defined by the finite element mesh shown in Figure 2.7 is given in Table 2.1.

Table 2.1 COMPARISON OF VALUES OF u FOR SOME INTERNAL POINTS DEFINED BY FINITE ELEMENT MESH SHOWN IN FIGURE 2.7

Node	*Finite element*	*Constant boundary element*	*Cell collocation*
16	1·154	1·227	1·230
27	0·801	0·824	0·829
44	0·404	0·426	0·416
46	1·205	1·217	1·210
58	1·205	1·214	1·210
63	0·768	0·776	0·771

2.5 POISSON'S EQUATION

In many practical problems we need to introduce body type forces into the formulation which are acting on the volume or area of the body but will not add any internal unknowns to the formulation.

The starting expression is given by equation (2.23), i.e.

$$\tfrac{1}{2}u_i + \int_\Omega b u^* \, d\Omega + \int_\Gamma u q^* \, d\Gamma = \int_\Gamma q u^* \, d\Gamma \tag{2.68}$$

In order to integrate the b term we need to discretise the domain Ω into a series of 'cells' or 'interior elements', similar to those used in the finite element method (Figure 2.10) but conceptually different as we will not require here any internal unknowns.

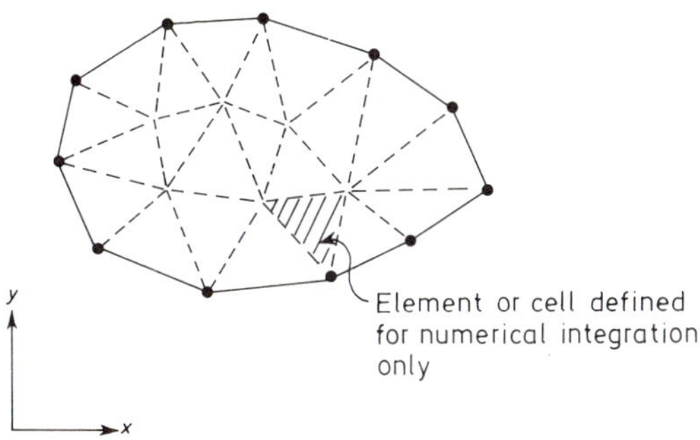

Figure 2.10 Boundary and internal cells

Consider M internal elements, we can then write,

$$B_i = \int_{\Omega} bu^* \mathrm{d}\Omega = \sum_{k=1}^{M} \left(\int_{\Gamma_k} bu^* \mathrm{d}\Gamma \right) \tag{2.69}$$

Over each element a numerical integration formula can then be applied, such as,

$$\int_{\Omega} bu^* \mathrm{d}\Omega = \sum_{k=1}^{M} \left(\sum_{r=1}^{S} w_r (bu^*)_r \right) A_k \tag{2.70}$$

where r is the integration point, w_r the weighting function, S the total number of integration points on each k cell and A_k the area of the cell.

Hence for each boundary point 'i' equation (2.68) can be written in discretised form as follows,

$$B_i + \sum_{j=1} H_{ij} u_j = \sum_{j=1}^{N} G_{ij} q_j \tag{2.71}$$

where the H_{ij} *and* G_{ij} terms are the same as discussed previously (Section 2.4).

The whole set of equations for the N nodes can be expressed in matrix form as follows,

$$\mathbf{B} + \mathbf{HU} = \mathbf{GQ} \tag{2.72}$$

Note that N_1 values of u and N_2 values of q are known on the boundary. Hence equations (2.72) are reordered in such a way that all the unknowns are on the left-hand side. This gives,

$$\mathbf{AX} = \mathbf{F} \tag{2.73}$$

where the $\mathbf{F}$ vector contains the terms of $\mathbf{B}$.

Once the values of u and q are known over the whole boundary we can calculate their values at any internal points taking into account the contribution of the b terms. For instance, the internal value of u at an internal point i is now given by,

$$u_i = \sum_{j=1}^{N} G_{ij} q_j - \sum_{j=1}^{N} \hat{H}_{ij} u_j - B_i \tag{2.74}$$

Similar considerations can be made for the source solution. Here we start with equations (2.35) and (2.36) including boundary terms, i.e.

$$u_i = \int_\Gamma \sigma u^* \, d\Gamma - \int_\Omega b u^* \, d\Omega$$
$$q_k = -\tfrac{1}{2}\sigma_k + \int_\Gamma \sigma q^* \, d\Gamma - \int_\Omega b q^* \, d\Omega \tag{2.75}$$

and integrate them numerically, which gives,

$$u_i = \sum_{j=1}^{N} \sigma_j G_{ij} - B_i^u$$
$$q_k = \sum_{j=1}^{N} \sigma_j H_{kj} - B_k^q \tag{2.76}$$

The terms B^u and B^q can be obtained applying numerical integration. Once the boundary conditions are applied, the system of equations (2.76) can be written in the same form as (2.73).

2.6 THE ORTHOTROPIC CASE

In many engineering applications the material properties cannot be considered to be isotropic and we need to consider them as orthotropic. Let us study the case of a body such as the one represented in Figure 2.11. The governing equation for this system,

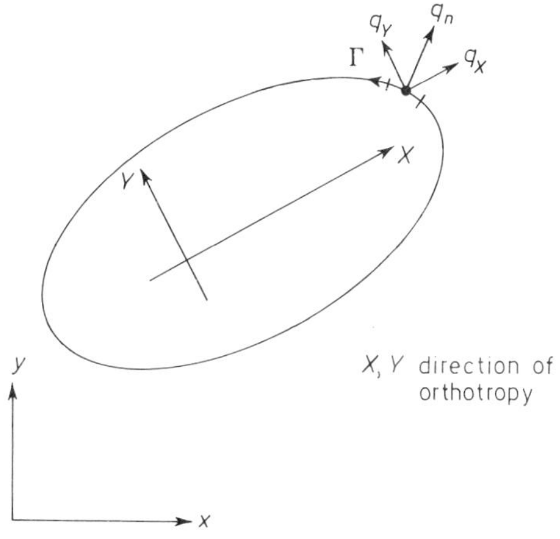

Figure 2.11 Orthotropic medium

referred to the X_1, X_2, X_3 axes of orthotropy can be written as,

$$\nabla_k^2 u = 0 \tag{2.77}$$

where for the three-dimensional case,

$$\nabla_k^2(\ ) = k_1 \frac{\partial^2(\ )}{\partial X_1} + k_2 \frac{\partial^2(\ )}{\partial X_2} + k_3 \frac{\partial^2(\ )}{\partial X_3} \tag{2.78}$$

or for two dimensions,

$$\nabla_k^2(\ ) = k_1 \frac{\partial^2(\ )}{\partial X_1} + k_2 \frac{\partial^2(\ )}{\partial X_2} \tag{2.79}$$

The k_i terms define the material properties in the directions of orthotropy.

If we now assume that a concentrated potential is acting at a point 'i', the fundamental solution should satisfy the following equation:

$$\nabla_k^2 u^* + \delta_i = 0 \tag{2.80}$$

where δ_i can be written for three dimensions as,

$$\delta(X_1 - X_{1i})\,\delta(X_2 - X_{2i})\,\delta(X_3 - X_{3i}) \tag{2.81}$$

The simplest way of finding the fundamental solution is to make the following transformation:

$$\xi_1 = \frac{X_1}{\sqrt{k_1}}, \qquad \xi_2 = \frac{X_2}{\sqrt{k_2}}, \qquad \xi_3 = \frac{X_3}{\sqrt{k_3}} \tag{2.82}$$

We must now use the property that

$$\delta(X_1 - X_{1i}) = \delta(\sqrt{k_1}\,(\xi_1 - \xi_{1i})) = \frac{1}{\sqrt{k_1}}\delta(\xi_1 - \xi_{1i}) \tag{2.83}$$

which can be deduced by considering the integral

$$\begin{aligned} &\int_{\xi_{1i}-a}^{\xi_{1i}+a} u(\xi_1)\,\delta(\sqrt{k_1}\,(\xi_1 - \xi_{1i}))\,\mathrm{d}\xi_1 \\ &\quad = \frac{1}{\sqrt{k_1}} \int_{X_{1i}-a\sqrt{k_1}}^{X_{1i}+a\sqrt{k_1}} u\left(\frac{X_1}{\sqrt{k_1}}\right) \delta(X_1 - X_{1i})\,\mathrm{d}X_1 \\ &\quad = \frac{1}{\sqrt{k_1}} u\left(\frac{X_{1i}}{\sqrt{k_1}}\right) = \frac{1}{\sqrt{k_1}} u(\xi_{1i}) \end{aligned} \tag{2.84}$$

where a is a positive constant. Hence equation (2.80) can be written

$$\nabla^2 u = \frac{\partial^2 u}{\partial \xi_1^2} + \frac{\partial^2 u}{\partial \xi_2^2} + \frac{\partial^2 u}{\partial \xi_3^2} + \frac{1}{\sqrt{k_1 k_2 k_3}} \delta(\xi_1 - \xi_{1i})\delta(\xi_2 - \xi_{2i})\delta(\xi_3 - \xi_{3i})$$

$$= 0 \tag{2.85}$$

The fundamental solution of equation (2.85) is as previously

$$u^* = \frac{1}{4\pi r_0} \frac{1}{\sqrt{k_1 k_2 k_3}} \tag{2.86}$$

where,

$$r_0 = \sqrt{\xi_1^2 + \xi_2^2 + \xi_3^2} = \sqrt{\frac{X_1^2}{k_1} + \frac{X_2^2}{k_2} + \frac{X_3^2}{k_3}} \tag{2.87}$$

The same transformation applied to two-dimensional problems produces,

$$u^* = \frac{1}{2\pi\sqrt{k_1 k_2}} \ln\left(\frac{1}{r_0}\right) \tag{2.88}$$

where now

$$r_0 = \sqrt{\frac{X_1^2}{k_1} + \frac{X_2^2}{k_2}}$$

FLUX AT THE BOUNDARY

Consider the two-dimensional case for simplicity. In order to calculate the flux at the boundary we can apply Green's theorem, which gives,

$$\iint_\Omega \left(k_1 \frac{\partial^2 u}{\partial X^2} + k_2 \frac{\partial^2 u}{\partial Y^2} \right) d\Omega \Rightarrow \int_\Gamma \left(k_1 \frac{\partial u}{\partial X} n_X + k_2 \frac{\partial u}{\partial Y} n_Y \right) d\Gamma \tag{2.89}$$

where

$$n_X = \cos(n, X), \qquad n_Y = \cos(n, Y)$$

The term between brackets in the right-hand side is the internal flux component on the boundary, i.e.

$$q = k_1 \frac{\partial u}{\partial X} n_X + k_2 \frac{\partial u}{\partial Y} n_Y \tag{2.90}$$

This expression allows us to calculate the value of q at any point on the boundary. The formula also applies when u is substituted by the fundamental solution u^*.

2.7 THE HELMHOLTZ EQUATION

Let us now consider the Helmholtz equation, i.e.

$$\nabla^2 u + \lambda^2 u = 0 \tag{2.91}$$

where λ^2 is a positive constant. The boundary conditions are assumed to be

$$\begin{aligned} u &= \bar{u} && \text{on } \Gamma_1 \\ \partial u/\partial n = q &= \bar{q} && \text{on } \Gamma_2 \end{aligned} \tag{2.92}$$

The starting weighted residual statement for this case is,

$$\int_\Omega (\nabla^2 u + \lambda^2 u) u^* \, d\Omega = \int_{\Gamma_2} (q - \bar{q}) u^* \, d\Gamma - \int_{\Gamma_1} (u - \bar{u}) q^* \, d\Gamma \tag{2.93}$$

Integrating twice by parts we obtain,

$$\int_\Omega (\nabla^2 u^* + \lambda^2 u^*) u \, d\Omega = -\int_\Gamma q u^* \, d\Gamma + \int_\Gamma q^* u \, d\Gamma \tag{2.94}$$

where $\Gamma = \Gamma_1 + \Gamma_2$. u^* is the fundamental solution for the Helmholtz equation, which satisfies

$$\nabla^2 u^* + \lambda^2 u^* + \delta_i = 0 \tag{2.95}$$

For two dimensions this fundamental solution is

$$u^* = +\frac{1}{4\mathrm{i}} H_0^{(2)}(\lambda r)$$

and (2.96)

$$\frac{\partial u^*}{\partial r} = -\frac{\lambda}{4\mathrm{i}} H_1^{(2)}(\lambda r)$$

r is the distance from the source point to the point under consideration. $H_0^{(2)}$ is a Hankel function of the second kind of zero order, i.e.

$$H_0^{(2)}(\lambda r) = J_0(\lambda r) - \mathrm{i}\, Y_0(\lambda r) \tag{2.97}$$

where J_0 is a Bessel function of the first kind and Y_0 is one of the second kind. The Hankel function of order one is

$$H_1^{(2)}(\lambda r) = J_1(\lambda r) - \mathrm{i}\, Y_1(\lambda r) \tag{2.98}$$

For three dimensions, the fundamental solution is,

$$u^* = \frac{1}{4\pi r} \mathrm{e}^{-\mathrm{i}\lambda r}$$

and (2.99)

$$\frac{\partial u^*}{\partial r} = \frac{1}{4\pi r}\left(\frac{1}{r} - \mathrm{i}\lambda\right)\mathrm{e}^{-\mathrm{i}\kappa r}$$

Substituting (2.93) into (2.92) we obtain,

$$u_i + \int_\Gamma u q^* \,\mathrm{d}\Gamma = \int_\Gamma q u^* \,\mathrm{d}\Gamma \tag{2.100}$$

If the point 'i' is taken to be on the smooth Γ boundary, we will have for two and three dimensions the following result:

$$\tfrac{1}{2} u_i + \int_\Gamma u q^* \,\mathrm{d}\Gamma = \int_\Gamma q u^* \,\mathrm{d}\Gamma \tag{2.101}$$

In general we can write a constant c_i instead of $\frac{1}{2}$ for a non-smooth boundary. Hence,

$$c_i u_i + \int_\Gamma u q^* \,\mathrm{d}\Gamma = \int_\Gamma q u^* \,\mathrm{d}\Gamma \tag{2.102}$$

From now on the usual boundary element matrix technique can be applied.

Let us now consider how the solution behaves at infinity by assuming that we have two boundaries, one a closed boundary and the other the one at infinity (Figure 2.12). Formula (2.102) can now be written as

$$c_i u_i + \int_\Gamma (u q^* - q u^*) \mathrm{d}\Gamma = \int_{\Gamma_\infty} (q u^* - q^* u)\, \mathrm{d}\Gamma \tag{2.103}$$

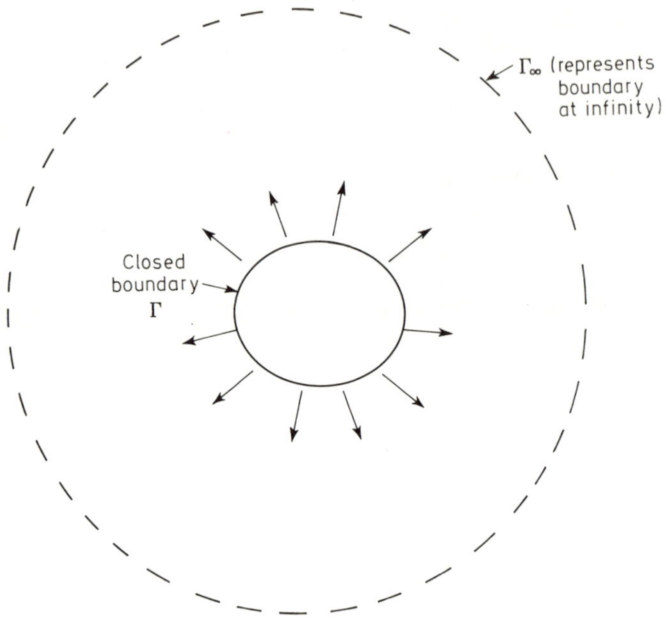

Figure 2.12 Representation of the region extending to infinity

Consider only the last term in (2.103) for the case of a three-dimensional medium, for which the boundary at Γ_∞ will be a sphere of radius tending to infinity. We can write

$$\int_{\Gamma_\infty} (qu^* - q^*u)\,\mathrm{d}\Gamma = \int_{\Gamma_\infty} -\frac{1}{4\pi r}\mathrm{e}^{-\mathrm{i}\lambda r}\left[q - \left(\frac{1}{r} - i\lambda\right)u\right] r^2\,\mathrm{d}\theta\mathrm{d}\phi \tag{2.104}$$

The minus sign in front of the right-hand side integral comes from the fact that the normal on Γ_∞ is in the opposite direction to the radial coordinate. $\mathrm{d}\theta$, $\mathrm{d}\phi$ are the differentials of the angles in spherical polar coordinates.

Consider what happens to (2.104) when $r \to \infty$, i.e. consider

$$\lim_{r\to\infty}\left(\int_{\Gamma_\infty} -\frac{1}{4\pi}\mathrm{e}^{-\mathrm{i}\lambda r}\left[(q + i\lambda u)r - u\right]\mathrm{d}\theta\,\mathrm{d}\phi\right) \equiv 0 \tag{2.105}$$

In order for this expression to be zero in the limit we need to have,

$$(q + i\lambda u)r \Rightarrow 0 \qquad \text{as } r \Rightarrow \infty \tag{2.106}$$

Notice that we assume the term in u will go to zero as $u \to 0$ when $r \to \infty$.

The above condition (2.106) is the Sommerfeld radiation condition[5] in three dimensions and can be written as,

$$r\left(\frac{\partial u}{\partial r} + i\lambda u\right) \Rightarrow 0 \qquad \text{as } r \Rightarrow \infty \tag{2.107}$$

For two dimensions the condition is

$$\sqrt{r}\left(\frac{\partial u}{\partial r} + i\lambda u\right) \Rightarrow 0 \tag{2.108}$$

These conditions are satisfied by the fundamental solution.

REFERENCES

1. Brebbia, C. A. and Butterfield, R., 'The Formal Equivalence of the Direct and Indirect Boundary Element Methods', *Appl. Math. Modelling*, **2**, No. 2, June (1978)
2. Stroud, A. H. and Secrest, D., *Gaussian Quadrature Formulae*, Prentice Hall, New York, (1966)
3. Martin, H. C. 'Finite Element Analysis of Fluid Flows', *Proc. Conf. on Matrix Methods in Structural Mechanics*, AFFDL TR 68–150, Patterson Air Force Base, Ohio, USA (1969)
4. Lau, P. and Brebbia, C. A., 'The Cell Collocation Method', *Int. J. Mech. Sci.*, **20,** 83–95 (1978)
5. Sommerfeld, A., *Partial Differential Equations in Physics* New York, Academic Press, pp. 256–344 (1949)

3 Higher-Order Elements

3.1 INTRODUCTION

It was assumed in Chapter 2 that u and q were constant over each element. In general however, u and q can have a linear or higher-order variation. In addition their variations need not be of the same order and for instance, it is more consistent to take q of one order less than u as it is equal to the derivative of the potential.

In this chapter we will start by discussing how elements with linear variations of u and q (i.e. linear elements) can be developed for two-dimensional problems. We then progress to quadratic and higher-order elements, for two-dimensional problems.

If the body is three-dimensional however, the boundary elements are part of the external surface of the body and they are usually of two types: quadrilateral and triangular elements. To describe these elements we need to pass from the global three-dimensional coordinate system to a two-dimensional system defined over the surface of the body. Different types of elements are presented in this section together with the necessary transformation laws.

Finally, three-dimensional elements, similar to classical finite elements, are discussed. These elements may be useful to represent parts of the body where integration over the domain is required (e.g. body force terms).

3.2 LINEAR ELEMENTS FOR TWO-DIMENSIONAL PROBLEMS

Let us consider here a linear variation for u and q for which case the nodes are now considered to be at the intersection between two straight elements such as those shown in Figure 2.4(b).

Equation (2.42) can now be written,

$$c_i u_i + \sum_{j=1}^{N} \int_{\Gamma_j} u\, q^* \,\mathrm{d}\Gamma = \sum_{j=1}^{N} \int_{\Gamma_j} q\, u^* \,\mathrm{d}\Gamma \tag{3.1}$$

Note that the $-\frac{1}{2}$ coefficient of u_i has been replaced by an unknown c_i value. This is because c_i is equal to $-\frac{1}{2}$ only for a smooth boundary. The determination of c_i is rather complex but it will soon be shown that it does not need to be known explicitly and this simplifies the formulation.

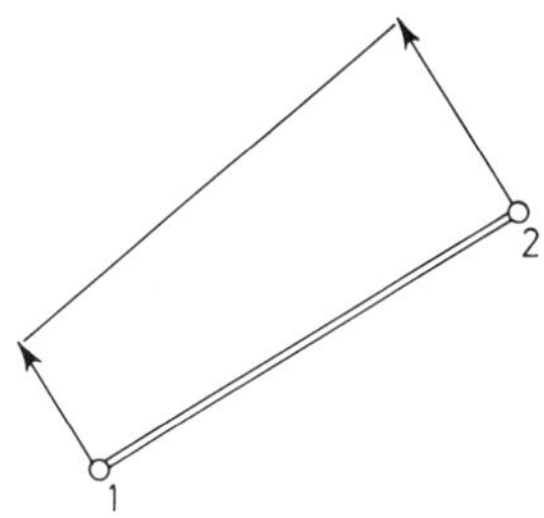

Figure 3.1 Linear element

The integrals in equation (3.1) are now more difficult to evaluate than in Section 2.5 because the u's and q's vary linearly over the element, in the way shown in Figure 3.1. The values of u and q at any point on the element can be defined in terms of their nodal values and two linear interpolation functions ϕ_1 and ϕ_2 functions of the homogeneous coordinate ξ and such that,

$$\begin{aligned} u(\xi) &= \phi_1 u_1 + \phi_2 u_2 = [\phi_1 \; \phi_2] \begin{Bmatrix} u_1 \\ u_2 \end{Bmatrix} \\ q(\xi) &= \phi_1 q_1 + \phi_2 q_2 = [\phi_1 \; \phi_2] \begin{Bmatrix} q_1 \\ q_2 \end{Bmatrix} \end{aligned} \tag{3.2}$$

The dimensionless coordinate ξ is equal to $x/\frac{1}{2}l$ and the ϕ_1, ϕ_2 functions are given by,

$$\phi_1 = \tfrac{1}{2}(1 - \xi), \qquad \phi_2 = \tfrac{1}{2}(1 + \xi) \tag{3.3}$$

The integrals along a segment j on the right-hand side of equation (3.1) can be written as,

$$\begin{aligned}\int_{\Gamma_j} u\,q^*\,\mathrm{d}\Gamma &= \int_{\Gamma_j} [\phi_1\,\phi_2]\,q^*\,\mathrm{d}\Gamma \begin{Bmatrix} u_1 \\ u_2 \end{Bmatrix} \\ &= [h_{ij}^1\ h_{ij}^2] \begin{Bmatrix} u_1 \\ u_2 \end{Bmatrix}\end{aligned} \tag{3.4}$$

where

$$h_{ij}^1 = \int_{\Gamma_j} \phi_1\,q^*\,\mathrm{d}\Gamma, \qquad h_{ij}^2 = \int_{\Gamma_j} \phi_2\,q^*\,\mathrm{d}\Gamma$$

The h_{ij}^k are influence coefficients defining the interaction between the point 'i' under consideration and a particular k node on an element 'j'.

For the integrals on the right-hand side of (3.1) we can write,

$$\begin{aligned}\int_{\Gamma_j} q\,u^*\,\mathrm{d}\Gamma &= \int_{\Gamma_j} [\phi_1\,\phi_2]\,u^*\,\mathrm{d}\Gamma \begin{Bmatrix} q_1 \\ q_2 \end{Bmatrix} \\ &= [g_{ij}^1\ g_{ij}^2] \begin{Bmatrix} q_1 \\ q_2 \end{Bmatrix}\end{aligned} \tag{3.5}$$

with,

$$g_{ij}^1 = \int_{\Gamma_j} \phi_1\,u^*\,\mathrm{d}\Gamma, \qquad g_{ij}^2 = \int_{\Gamma_j} \phi_2\,u^*\,\mathrm{d}\Gamma$$

To write the equation corresponding to node 'i' in discrete form we need to add the contribution from two adjoining elements, $(j-1)$ and (j), into one term, defining the nodal coefficient. This will give the following equation:

$$c_i u_i + [\hat{H}_{i1}\,\hat{H}_{i2}\,\ldots\,\hat{H}_{iN}] \begin{Bmatrix} u_1 \\ u_2 \\ \vdots \\ u_N \end{Bmatrix} = [G_{i1}\,G_{i2}\ldots G_{iN}] \begin{Bmatrix} q_1 \\ q_2 \\ \vdots \\ q_N \end{Bmatrix} \tag{3.6}$$

where each $\hat{H}_{ij}$ term is equal to the h^2 term of element '$j-1$' plus the h^1 term of element $(j-1)$ for a clockwise numbering system. The same applies for G_{ij}. Hence formula (3.6) represents the assembled equation for node 'i' and it can be written as,

$$c_i u_i + \sum_{j=1}^{N} \hat{H}_{ij} u_j = \sum_{j=1}^{N} G_{ij} q_j \tag{3.7}$$

or more simply,

$$\sum_{j=1}^{N} H_{ij} u_j = \sum_{j=1}^{N} G_{ij} q_j \tag{3.8}$$

where

$$H_{ij} = \hat{H}_{ij} \qquad \text{for } i \neq j$$

$$H_{ij} = \hat{H}_{ij} + c_i \quad \text{for } i = j$$

When all the nodes are taken into consideration, equation (3.8) produces an $N \times N$ system of equations which can be represented in matrix form as follows,

$$\mathbf{H}\,\mathbf{U} = \mathbf{G}\,\mathbf{Q} \tag{3.9}$$

If the surface is not smooth at the point 'i' the $c_i = -\frac{1}{2}$ is no longer valid but one can calculate the diagonal terms of $\mathbf{H}$ by using the fact that when a uniform potential is applied on the whole boundary the values of q's must be zero. Under this condition equation (3.9) produces,

$$\mathbf{H}\,\mathbf{I} = \mathbf{0} \tag{3.10}$$

Equation (3.10) indicates that the sum of all the elements of $\mathbf{H}$ in a row ought to be zero, hence the values of the coefficients in the diagonal can be easily calculated once the off-diagonal coefficients are all known, i.e.

$$H_{ii} = -\sum_{i \neq j} H_{ij} \tag{3.11}$$

Example 3.1

The Saint Venant torsion of a prismatic beam is governed by a differential equation of the type

$$\frac{\partial}{\partial x}\left(\frac{1}{G}\frac{\partial u}{\partial x}\right) + \frac{\partial}{\partial y}\left(\frac{1}{G}\frac{\partial u}{\partial y}\right) = -2\theta \tag{a}$$

where G is the shear modulus and θ is the rate of twist. The problem variable is the stress function u, such that

$$\tau_{xz} = \frac{\partial u}{\partial y}, \qquad \tau_{yz} = -\frac{\partial u}{\partial x} \tag{b}$$

τ's are the shear stresses. u must have a constant value on the boundary. For convenience it is normally assumed that such constant is equal to zero. Considering an homogeneous material we can write,

$$\frac{\partial^2 v}{\partial x^2} + \frac{\partial^2 v}{\partial y^2} = 2 \tag{c}$$

where $v = u/G\theta$.

This problem can be solved for v, but noticing that the rate of twist θ is so far unknown. We know that the torque M_t is

$$M_t = JG\theta = 2\int_\Omega u \, dx dy \tag{d}$$

where J is the torsional rigidity. We can therefore compute J by evaluating the following integral:

$$J = 2\int_\Omega v \, dx dy \tag{e}$$

The rate of twist is then given by

$$\theta = M_t/GJ \tag{f}$$

Finally we can compute the shear stresses as

$$\tau_{xz} = G\theta \frac{\partial v}{\partial y}, \qquad \tau_{yz} = -G\theta \frac{\partial v}{\partial x} \tag{g}$$

Consider as an illustration the problem of the torsion of a prismatic bar of elliptical cross section, defined by the equation,

$$\frac{x^2}{a} + \frac{y^2}{b} = 1 \tag{h}$$

For this example we take $a = 2$ and $b = 1$. A finite element mesh selected, consisting of 33 nodes and 48 linear elements is shown in Figure 3.2.

The same example was solved using linear boundary elements (Figure 3.3) with 16 nodes and a series of internal points coinciding

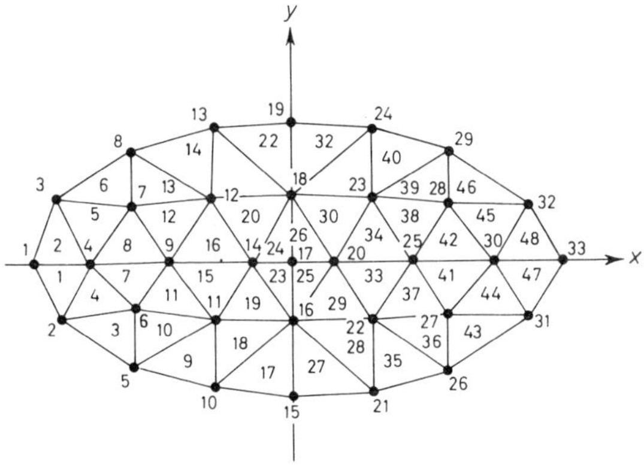

Figure 3.2 Discretisation of the elliptical section using finite elements

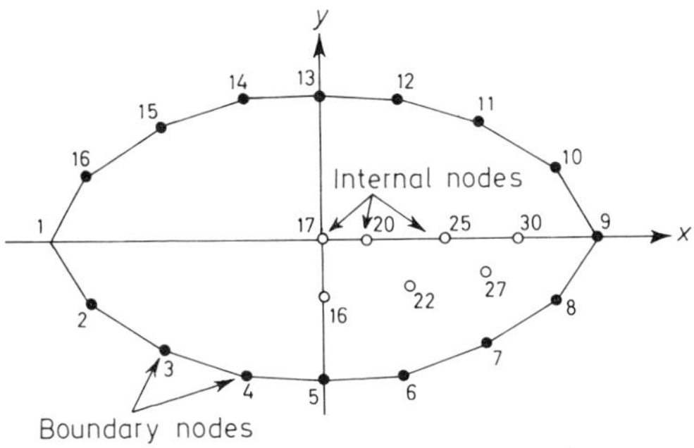

Figure 3.3 Discretisation of the section using boundary elements

with the internal finite element nodes.

The analytical solution for this problem gives $J = 5{\cdot}026$ for the torsional rigidity.

Taking $G = \theta = 1$, one can compute

$$M_t = G\theta J = 5{\cdot}026 \tag{i}$$

Table 3.1 includes a comparison between analytical and computed values for the problem variable and the torsional rigidities.

Having the value of the integral of the problem variable over the area of the ellipse, one can compute the approximate value of the torsional rigidity J obtaining

$$J = 2\int_{\Omega} v\,\mathrm{d}\Omega = \begin{cases} 4{\cdot}560 & \text{for finite elements} \\ 4{\cdot}487 & \text{for boundary elements} \end{cases}$$

The approximate solutions are about 10 % of the analytical values. Notice that better agreement is obtained for the problem variable than for the torsional rigidity, which is computed approximately from the approximate values of the problem variables.

Table 3.1 VALUE OF TORSIONAL RIGIDITY AND v FUNCTION FOR INTERNAL POINTS

Y	*Exact*	*Finite elements (linear)*	*Boundary elements (linear)*
0	0·350	0·341	0·334
−0·35	0·414	0·392	0·401
0	0·638	0·627	0·626
−0·75	0·566	0·561	0·557
0	0·782	0·790	0·772
−0·75	0·638	0·665	0·629
0	0·800	0·793	0·791
Rigidity	5·026	4·560	4·487

Example 3.2

Figure 3.4 describes the aerofoil shape called NACA 0018 which can be studied using boundary elements. The surface of the aerofoil was

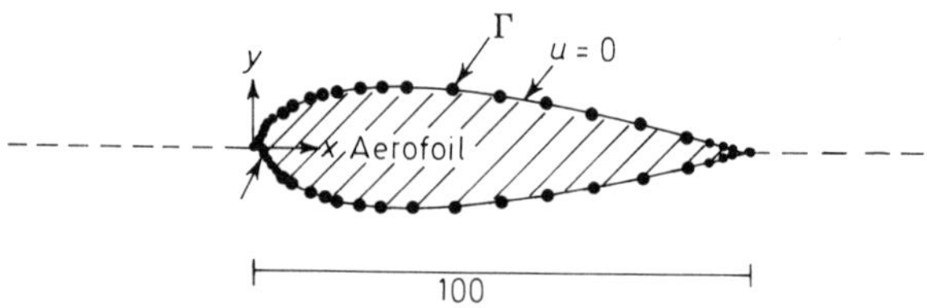

Figure 3.4 NACA 0018 aerofoil

divided into a series of elements. Because of symmetry only half the aerofoil needs to be considered. The problem can be expressed in terms of a stream function u for a perfect fluid, such that the velocities are

$$v_x = \frac{\partial u}{\partial y}, \qquad v_y = -\frac{\partial u}{\partial x} \tag{a}$$

The velocities far from the aerofoil are,

$$v_x = V, \qquad v_y = 0 \tag{b}$$

The solution is considered to be the superposition of two solutions, u_1 for the flow without the aerofoil and u_2 for the disturbed flow around the aerofoil,

$$u = u_1 + u_2 \tag{c}$$

where u_1 produces the velocity field defined by (b). Considering $u = 0$ on the aerofoil equation (c) gives

$$u_1 = -u_2$$

for all the points on the surface. The results for the tangential velocity are shown in Table 3.2 where they are compared against those presented by NACA. The agreement is satisfactory.

Table 3.2 COMPARISON OF BOUNDARY ELEMENTS AND ANALYTICAL SOLUTIONS FOR THE TANGENTIAL VELOCITIES NACA 0018 AEROFOIL

	Solution for (v/V)	
x	*Analytical*	*Boundary element*
0·0	0·000	0·000
1·25	0·926	0·931
2·5	1·103	1·093
5·0	1·228	1·212
7·5	1·267	1·253
10.0	1·276	1·265
15·0	1·278	1·273
20·0	1·275	1·269
25·0	1·262	1·254
30·0	1·247	1·236
40·0	1·205	1·198
50·0	1·157	1·158
60·0	1·116	1·118
70·0	1·074	1·079
80·0	1·025	1·036
90·0	0·966	0·975
95·0	0·914	0·941
100·0	0·000	0·000

v is the tangential velocity.

3.3 QUADRATIC AND HIGHER-ORDER ELEMENTS

Quadratic elements are also frequently used to represent better the geometry of the body. They do not present any special difficulties, but we need to transform from Cartesian to curvilinear coordinates.

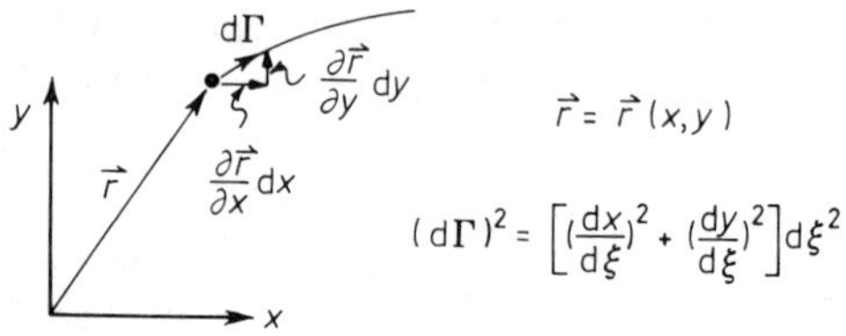

Figure 3.5 Curved boundary notation

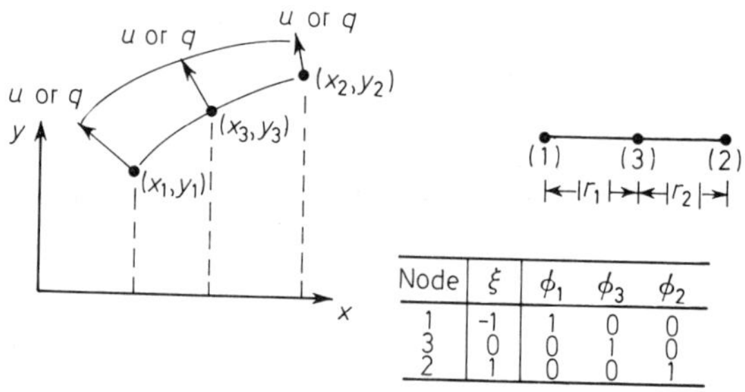

Node	ξ	ϕ_1	ϕ_3	ϕ_2
1	-1	1	0	0
3	0	0	1	0
2	1	0	0	1

Figure 3.6 Quadratic element

Consider the curved boundary depicted in Figure 3.5 and the element shown in Figure 3.6. The u and q functions can be written in terms of the homogeneous ξ coordinate as follows,

$$u(\xi) = \phi_1 u_1 + \phi_2 u_2 + \phi_3 u_3 = [\phi_1 \, \phi_2 \, \phi_3] \begin{Bmatrix} u_1 \\ u_2 \\ u_3 \end{Bmatrix}$$

$$q(\xi) = \phi_1 q_1 + \phi_2 q_2 + \phi_3 q_3 = [\phi_1 \, \phi_2 \, \phi_3] \begin{Bmatrix} q_1 \\ q_2 \\ q_3 \end{Bmatrix} \tag{3.12}$$

where,

$$\phi_1 = \tfrac{1}{2}\xi(\xi - 1), \qquad \phi_2 = \tfrac{1}{2}\xi(1 + \xi), \qquad \phi_3 = (1 - \xi)(1 + \xi)$$

Note that these functions are such that they give the nodal values of the variable under consideration when specialised for the nodes (see table in Figure 3.6) and that they vary quadratically.

The integrals along a '*j*' element in equation (1) are similar to those for a linear element. The integral for u for instance is,

$$\int_{\Gamma_j} u(\xi) q^* \,\mathrm{d}\Gamma = \int_{\Gamma_j} [\phi_1 \, \phi_2 \, \phi_3] \, q^* \,\mathrm{d}\Gamma \begin{Bmatrix} u_1 \\ u_2 \\ u_3 \end{Bmatrix}$$

$$= [h_{ij}^1 \; h_{ij}^2 \; h_{ij}^3] \begin{Bmatrix} u_1 \\ u_2 \\ u_3 \end{Bmatrix} \tag{3.13}$$

and similarly for q.

The evaluation of the integrals requires the use of a Jacobian as ϕ_i are functions of ξ but the integrals are with respect to Γ. For a two-dimensional such as this the transformation is simple as the Jacobian becomes,

$$|G| = \sqrt{\left(\frac{\mathrm{d}x}{\mathrm{d}\xi}\right)^2 + \left(\frac{\mathrm{d}y}{\mathrm{d}\xi}\right)^2} = \frac{\mathrm{d}\Gamma}{\mathrm{d}\xi} \tag{3.14}$$

Thus,

$$\mathrm{d}\Gamma = |G| \,\mathrm{d}\xi \tag{3.15}$$

Substituting this relationship into (3.13) we obtain,

$$\int_{\Gamma_j} u(\xi) q^* \,\mathrm{d}\Gamma = \int_{(1)}^{(2)} u(\xi) q^* |G| \,\mathrm{d}\xi \tag{3.16}$$

which can now be integrated numerically. Similar considerations apply for the integrals for q.

Note that to calculate (3.14) we need to express the variations of the x and y coordinates on the boundary in terms of ξ. This can be done by writing them in the same way as the functions for u and q where expressed, i.e.

$$x = \phi_1 x_1 + \phi_2 x_2 + \phi_3 x_3$$
$$y = \phi_1 y_1 + \phi_2 y_2 + \phi_3 y_3 \tag{3.17}$$

x_i, y_i are the coordinates of a node referred to the global system (see Figure 3.5).

It will be shown that in order to be able to reproduce the constant potential type conditions, the order of the u function has to be at least the same as the function for x and u used to describe the geometry of the body.

Higher-order elements can also be generated. For instance a cubic variation for u or q can be obtained by taking four nodes over each element (Figure 3.7). Here,

$$u = \phi_1 u_1 + \phi_2 u_2 + \phi_3 u_3 + \phi_4 u_4 \tag{3.18}$$

where

$$\phi_1 = \tfrac{1}{16}(1-\xi)[-10+9(\xi^2+1)] \quad \phi_2 = \tfrac{1}{16}(1+\xi)[-10+ +9(\xi^2+1)]$$
$$\phi_3 = \tfrac{9}{16}(1-\xi^2)(1-3\xi) \quad \phi_4 = \tfrac{9}{16}(1-\xi^2)(1+3\xi)$$

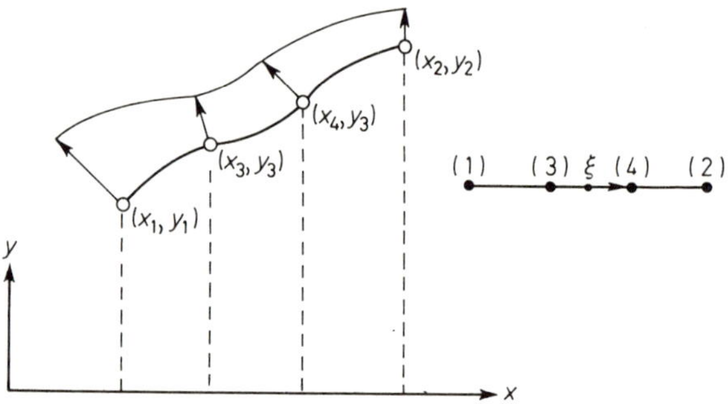

Node	ξ	ϕ_1	ϕ_2	ϕ_3	ϕ_4
1	-1	1	0	0	0
2	$-\frac{1}{3}$	0	1	0	0
3	$\frac{1}{3}$	0	0	1	0
4	1	0	0	0	1

Figure 3.7 Cubic elements

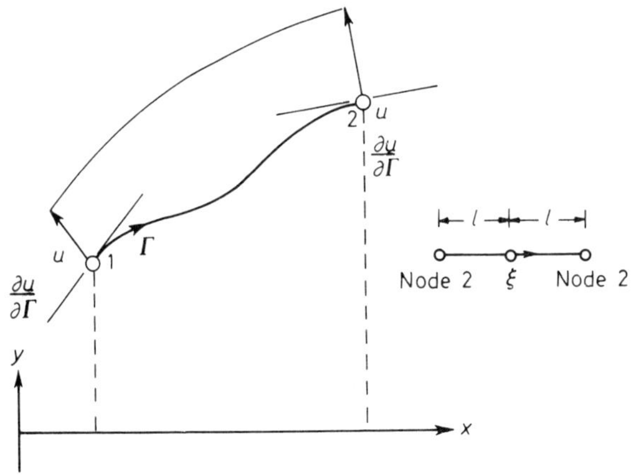

Figure 3.8 Cubic elements

We can also define the cubic variation of the function u (the same applies for q and the x, y coordinates) taking as unknowns the function and its derivative at the two extreme nodes (Figure 3.7). For this case,

$$u = \phi_1 u_1 + \phi_2 \left(\frac{\partial u}{\partial \Gamma}\right)_1 + \phi_3 u_2 + \phi_4 \left(\frac{\partial u}{\partial \Gamma}\right)_2 \tag{3.19}$$

and,

$$\phi_1 = \tfrac{1}{4}(\xi - 1)^2 (\xi + 2) \qquad \phi_2 = -\tfrac{1}{4} l (\xi - 1)^2 (\xi + 1)$$

$$\phi_3 = \tfrac{1}{4}(\xi + 1)^2 (\xi - 2) \qquad \phi_4 = -\tfrac{1}{4} l (\xi + 1)^2 (\xi - 1)$$

3.4 ELEMENTS FOR THREE-DIMENSIONAL PROBLEMS

If the body is three dimensional the boundary elements are part of the external surface of the body (Figure 3.9). They are usually of two types: quadrilateral and triangular elements. The functions used to describe their geometry or the variables u and q over them are similar to those used in finite elements.

In order to study these elements we need first to consider the way in which we can pass from the x, y, z global system to the system ξ_1, ξ_2, η defined over the surface of the body.

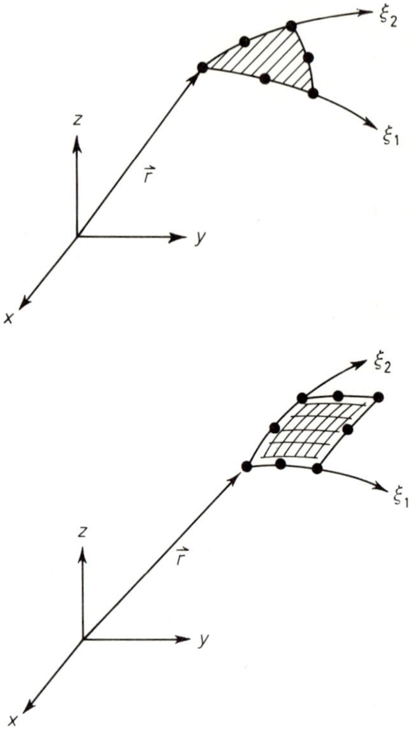

Figure 3.9 Elements for three-dimensional bodies

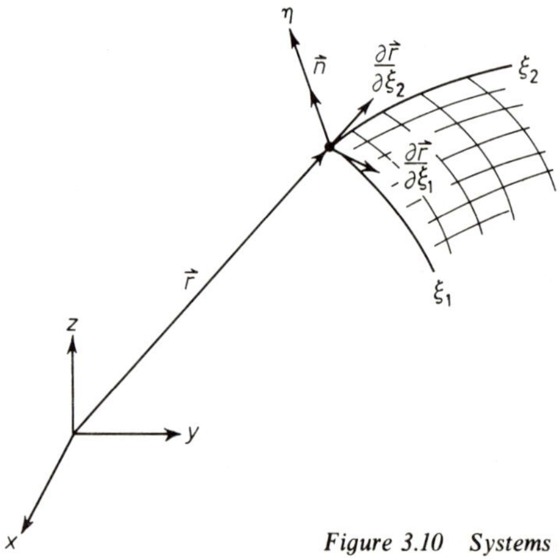

Figure 3.10 Systems of co-ordinates

Consider the systems defined in Figure 3.10. For a function u, the general transformation is given by

$$\begin{Bmatrix} \dfrac{\partial u}{\partial \xi_1} \\ \dfrac{\partial u}{\partial \xi_2} \\ \dfrac{\partial u}{\partial \eta} \end{Bmatrix} = \begin{bmatrix} \dfrac{\partial x}{\partial \xi_1} & \dfrac{\partial y}{\partial \xi_1} & \dfrac{\partial z}{\partial \xi_1} \\ \dfrac{\partial x}{\partial \xi_2} & \dfrac{\partial y}{\partial \xi_2} & \dfrac{\partial z}{\partial \xi_2} \\ \dfrac{\partial x}{\partial \eta} & \dfrac{\partial y}{\partial \eta} & \dfrac{\partial z}{\partial \eta} \end{bmatrix} \begin{Bmatrix} \dfrac{\partial u}{\partial x} \\ \dfrac{\partial u}{\partial y} \\ \dfrac{\partial u}{\partial z} \end{Bmatrix} \tag{3.20}$$

or,

$$\begin{Bmatrix} \dfrac{\partial u}{\partial \xi_1} \\ \dfrac{\partial u}{\partial \xi_2} \\ \dfrac{\partial u}{\partial \eta} \end{Bmatrix} = \mathbf{J} \begin{Bmatrix} \dfrac{\partial u}{\partial x} \\ \dfrac{\partial u}{\partial y} \\ \dfrac{\partial u}{\partial z} \end{Bmatrix} \tag{3.21}$$

Hence we can find the inverse relationship, i.e.

$$\begin{Bmatrix} \dfrac{\partial u}{\partial x} \\ \dfrac{\partial u}{\partial y} \\ \dfrac{\partial u}{\partial z} \end{Bmatrix} = \mathbf{J}^{-1} \begin{Bmatrix} \dfrac{\partial u}{\partial \xi_1} \\ \dfrac{\partial u}{\partial \xi_2} \\ \dfrac{\partial u}{\partial \eta} \end{Bmatrix} \tag{3.22}$$

Having defined these relationships we can take for instance $\vec{r}$ instead of u and define,

$$\begin{aligned} \mathrm{d}\Omega = \mathrm{d}\ (\text{volume}) &= \text{magnitude of} \left(\frac{\partial \vec{r}}{\partial \xi_1} \times \frac{\partial \vec{r}}{\partial \xi_2} \cdot \frac{\partial \vec{r}}{\partial \eta} \right) \mathrm{d}\xi_1 \, \mathrm{d}\xi_2 \, \mathrm{d}\eta \\ &= (\text{absolute value of } |\mathbf{J}|)\, \mathrm{d}\xi_1 \, \mathrm{d}\xi_2 \, \mathrm{d}\eta \end{aligned} \tag{3.23}$$

A differential of area (such as $\mathrm{d}\Omega$) for instance will be given by

$$\mathrm{d}(\text{area}) = \left| \frac{\partial \vec{r}}{\partial \xi_1} \times \frac{\partial \vec{r}}{\partial \xi_2} \right| \mathrm{d}\xi_1 \, \mathrm{d}\xi_2 = |G| \, \mathrm{d}\xi_1 \, \mathrm{d}\xi_2 \tag{3.24}$$

where

$$\frac{\partial \vec{r}}{\partial \xi_1} = \left(\frac{\partial x}{\partial \xi_1}, \frac{\partial y}{\partial \xi_1}, \frac{\partial z}{\partial \xi_1} \right)$$
$$\frac{\partial \vec{r}}{\partial \xi_2} = \left(\frac{\partial x}{\partial \xi_2}, \frac{\partial y}{\partial \xi_2}, \frac{\partial z}{\partial \xi_2} \right) \qquad (3.25)$$

Note that $|G|$ is the magnitude of the normal vector,

$$\vec{n} = \frac{\partial \vec{r}}{\partial \xi_1} \times \frac{\partial \vec{r}}{\partial \xi_2} = \left(\frac{\partial x}{\partial \eta}, \frac{\partial y}{\partial \eta}, \frac{\partial z}{\partial \eta} \right)$$
$$= (g_1, g_2, g_3) \qquad (3.26)$$

where

$$g_1 = \left(\frac{\partial y}{\partial \xi_1} \frac{\partial z}{\partial \xi_2} - \frac{\partial y}{\partial \xi_2} \frac{\partial z}{\partial \xi_1} \right)$$
$$g_2 = \left(\frac{\partial z}{\partial \xi_1} \frac{\partial x}{\partial \xi_2} - \frac{\partial x}{\partial \xi_1} \frac{\partial z}{\partial \xi_2} \right) \qquad (3.27)$$
$$g_3 = \left(\frac{\partial x}{\partial \xi_1} \frac{\partial y}{\partial \xi_2} - \frac{\partial y}{\partial \xi_1} \frac{\partial x}{\partial \xi_2} \right)$$

The magnitude of $\vec{n}$ gives the value of $|G|$, i.e.

$$|G| = (g_1^2 + g_2^2 + g_3^2)^{1/2} \qquad (3.28)$$

These relationships will be used to integrate terms such as

$$\int u^* q \, \mathrm{d}\Gamma \quad \text{and} \quad \int u q^* \, \mathrm{d}\Gamma \qquad (3.29)$$

which now become,

$$\int u^* q |G| \mathrm{d}\xi_1 \mathrm{d}\xi_2; \quad \int u q^* |G| \mathrm{d}\xi_1 \mathrm{d}\xi_2 \qquad (3.30)$$

The body force integrals like

$$\int b u^* \mathrm{d}\Omega \qquad (3.31)$$

will now become,

$$\int bu^* |J| \mathrm{d}\xi_1 \mathrm{d}\xi_2 \mathrm{d}\eta \tag{3.32}$$

u, b and q are assumed to be functions of ξ_1, ξ_2 and η where applicable.

QUADRILATERAL ELEMENTS

The simplest quadrilateral element is defined by its four corner points (Figure 3.11). We can use a set of dimensionless coordinates ξ_1, ξ_2 such that they vary from -1 to $+1$ and the interpolation functions ϕ_i. This allows us to write,

$$u = \phi_1 u_1 + \phi_2 u_2 + \phi_3 u_3 + \phi_4 u_4 \tag{3.33}$$

where the ϕ_i are,

$$\phi_1 = \tfrac{1}{4}(1-\xi_1)(1-\xi_2), \qquad \phi_2 = \tfrac{1}{4}(1+\xi_1)(1-\xi_2)$$

$$\phi_3 = \tfrac{1}{4}(1+\xi_1)(1+\xi_2), \qquad \phi_4 = \tfrac{1}{4}(1-\xi_1)(1+\xi_2)$$

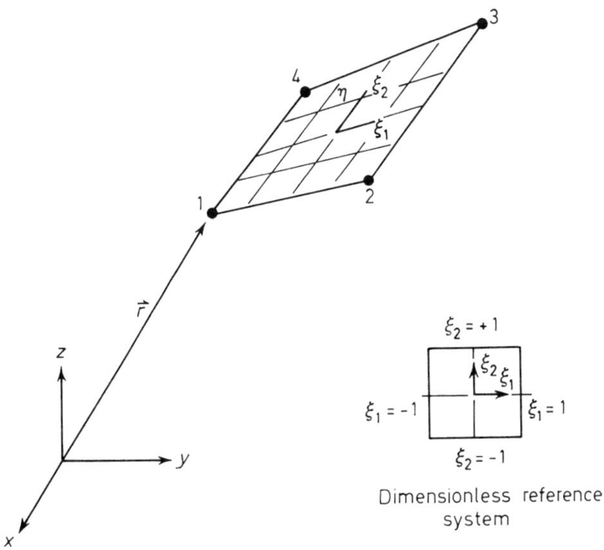

Figure 3.11 Simplest quadrilateral element

and u_i are the potentials at the nodes 1 to 4. The same function can be used to describe the geometry of the element, i.e.

$$x = \phi_1 x_1 + \phi_2 x_2 + \phi_3 x_3 + \phi_4 x_4 = \sum_{i=1}^{4} \phi_i x_i \tag{3.34}$$

and similarly for y and z.

Note that once these functions are defined we can obtain the value of $|G|$ and find the derivatives of a function such as u with respect to x, y or z using formula (3.22).

HIGHER-ORDER QUADRILATERAL ELEMENTS

We can improve our approximation by taking a second-order quadrilateral surface element (Figure 3.12), such that,

$$x = \sum_{i=1}^{8} \phi_i x_i \qquad y = \sum_{i=1}^{8} \phi_i y_i \qquad z = \sum_{i=1}^{8} \phi_i z_i \tag{3.35}$$

with,

$$\begin{aligned}
\phi_1 &= \tfrac{1}{4}(1-\xi_1)(1-\xi_2)(-\xi_1-\xi_2-1) \\
\phi_2 &= \tfrac{1}{4}(1+\xi_1)(1-\xi_2)(\xi_1-\xi_2-1) \\
\phi_3 &= \tfrac{1}{4}(1+\xi_1)(1+\xi_2)(\xi_1+\xi_2-1) \\
\phi_4 &= \tfrac{1}{4}(1-\xi_1)(1+\xi_2)(-\xi_1+\xi_2-1) \\
\phi_5 &= \tfrac{1}{2}(1-\xi_1^2)(1-\xi_2) \\
\phi_6 &= \tfrac{1}{2}(1-\xi_1^2)(1+\xi_1) \\
\phi_7 &= \tfrac{1}{2}(1-\xi_1^2)(1+\xi_2) \\
\phi_8 &= \tfrac{1}{2}(1-\xi_2^2)(1-\xi_1)
\end{aligned} \tag{3.36}$$

The same type of function could be used for u or q.

Higher-order quadrilateral elements can be used as boundary elements, such as the cubic type (cubic variation of the function on the element side), shown in Figure 3.13(a). For this element one has two additional nodes on each side, but one can also work with only corner nodes and include the derivatives as nodal quantities (Figure 3.13(b)).

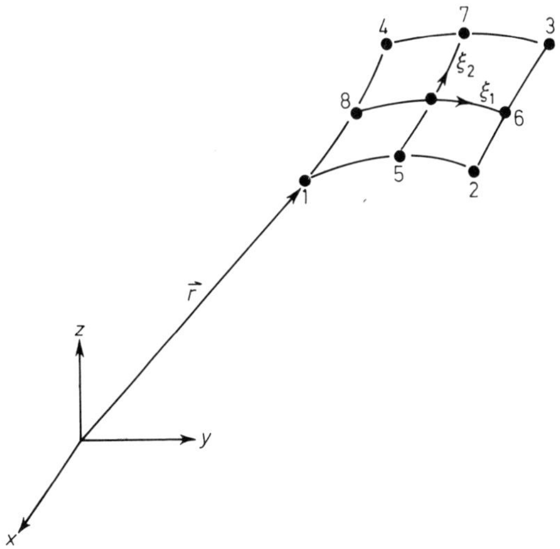

Figure 3.12 Eight-noded elements

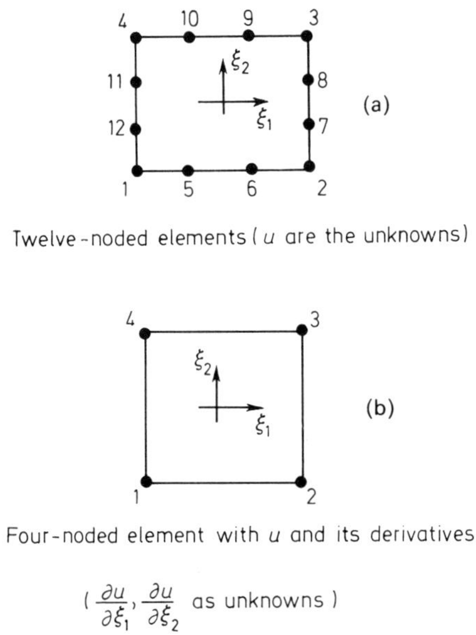

Figure 3.13 Higher-order elements

TRIANGULAR ELEMENTS

The simplest triangular element is one as shown in Figure 3.14. There is a linear variation of the function over the element and one can work with oblique coordinates ξ_1, ξ_2, which vary from 0 to 1 over the two sides shown in the figure.

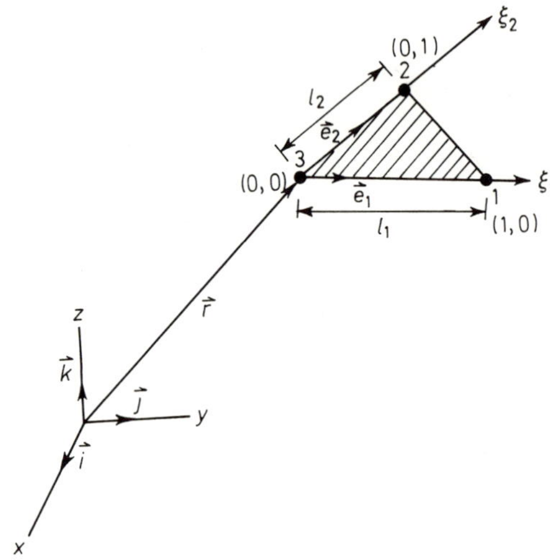

Figure 3.14 Simplest triangular element

The position vector $\vec{r}$ of any point on the triangle is,

$$\begin{aligned}\vec{r} &= x\vec{i} + y\vec{j} + z\vec{k} \\ &= x_3\vec{i} + y_3\vec{j} + z_3\vec{k} + l_1\xi_1\vec{e}_1 + l_2\xi_2\vec{e}_2\end{aligned} \tag{3.37}$$

where

$$\begin{aligned}\vec{e}_1 &= \left(\frac{x_1 - x_3}{l_1}\right)\vec{i} + \left(\frac{y_1 - y_3}{l_1}\right)\vec{j} + \left(\frac{z_1 - z_3}{l_1}\right)\vec{k} \\ \vec{e}_2 &= \left(\frac{x_1 - x_3}{l_2}\right)\vec{i} + \left(\frac{y_2 - y_3}{l_2}\right)\vec{j} + \left(\frac{z_2 - z_3}{l_2}\right)\vec{k}\end{aligned} \tag{3.38}$$

Hence the position vector $\vec{r}$ can be defined as,

$$\begin{aligned}\vec{r} &= [x_3 + (x_1 - x_3)\xi_1 + (x_2 - x_3)\xi_2]\vec{i} + \\ &\quad + [y_3 + (y_1 - y_3)\xi_1 + (y_2 - y_3)\xi_2]\vec{j} + \\ &\quad + [z_3 + (z_1 - z_3)\xi_1 + (z_2 - z_3)\xi_2]\vec{k} \\ &= x\vec{i} + y\vec{j} + z\vec{k}\end{aligned} \tag{3.39}$$

Thus we have that,

$$\begin{aligned}x &= \xi_1 x_1 + \xi_2 x_2 + (1 - \xi_1 - \xi_2)x_3 \\ y &= \xi_1 y_1 + \xi_2 y_2 + (1 - \xi_1 - \xi_2)y_3 \\ z &= \xi_1 z_1 + \xi_2 z_2 + (1 - \xi_1 - \xi_2)z_3\end{aligned} \tag{3.40}$$

We can now define $\xi_3 = 1 - \xi_1 - \xi_2$, where ξ_3 is a new but not independent coordinate, such that,

$$\xi_1 + \xi_2 + \xi_3 = 1$$

Hence,

$$\begin{Bmatrix} x \\ y \\ z \end{Bmatrix} = \begin{bmatrix} x_1 & x_2 & x_3 \\ y_1 & y_2 & y_3 \\ z_1 & z_2 & z_3 \end{bmatrix} \begin{Bmatrix} \xi_1 \\ \xi_2 \\ \xi_3 \end{Bmatrix} \tag{3.41}$$

As ξ_3 is not independent we can invert the first two equations to obtain,

$$\begin{aligned}\xi_1 &= \frac{1}{2A}(2A_1^0 + b_1 x + a_1 y) \\ \xi_2 &= \frac{1}{2A}(2A_2^0 + b_2 x + a_2 y)\end{aligned} \tag{3.42}$$

and obtain ξ_3 as $1 - \xi_1 - \xi_2$, i.e.

$$\xi_3 = \frac{1}{2A}(2A_3^0 + b_3 x + a_3 y) \tag{3.43}$$

The recurrence relationship for equations (3.42) and (3.43) is

$$a_i = x_k - x_j, \qquad b_i = y_j - y_k, \qquad 2A_i^0 = x_j y_k - x_k y_j \tag{3.44}$$

$i = 1, 2, 3$; $j = 2, 3, 1$; $k = 3, 1, 2$. The area A is

$$A = \tfrac{1}{2}(b_1 a_2 - b_2 a_1) \tag{3.45}$$

and represents the area of the projection of the element on the x, y plane.

The transformation can now be carried out as previously. The reason for using ξ_3 as a dependent variable is that it renders a simple expression for the interpolation functions but care should be taken not to confuse it with an independent variable.

Linear expressions for the coordinates or as u functions are now,

$$\begin{aligned} u &= u_1\xi_1 + u_2\xi_2 + u_3\xi_3 = \sum_{i=1}^{3} u_i\phi_i \\ x &= x_1\xi_1 + x_2\xi_2 + x_3\xi_3 = \sum_{i=1}^{3} x_i\phi_i \end{aligned} \tag{3.46}$$

Note that the interpolation functions are $\phi_i = \xi_i$.

The homogeneous triangular coordinates have one more advantage. This is that the integrals for polynomial terms can be carried out using a simple rule, i.e.

$$\iint \xi_1^i \xi_2^j \xi_3^k \, \mathrm{d}A = \frac{i!\,j!\,k!}{(i+j+k+2)!} 2A \tag{3.47}$$

This formula greatly simplifies the algebra when numerical integration is not used.

HIGHER-ORDER TRIANGULAR ELEMENTS

The next triangular model expresses the function as a complete second degree polynomial, which has a mid-side node (Figure 3.15). The model satisfies interelement compatibility and gives,

$$u = \sum \phi_i u_i \tag{3.48}$$

where

$$\begin{aligned} \phi_1 &= \xi_1(2\xi_1 - 1), & \phi_4 &= 4\xi_1\xi_2 \\ \phi_2 &= \xi_2(2\xi_2 - 1), & \phi_5 &= 4\xi_2\xi_3 \\ \phi_3 &= \xi_3(2\xi_3 - 1), & \phi_6 &= 4\xi_3\xi_1 \end{aligned} \tag{3.49}$$

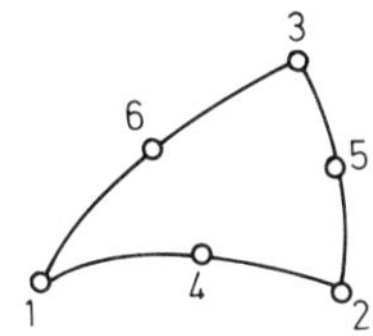

Figure 3.15 Quadratic element

The functions can also be expressed as complete cubic polynomials. Each component now involves ten parameters. In order to satisfy interelement compatibility the function for a side (which is cubic) must be represented with four nodal quantities. One possibility is to work with cornerpoint nodes and two interior nodes per side (Figure 3.16(a)), taking as nodal quantities, the u or q components. An additional interior node is needed to maintain completeness of the polynomial, since if the polynomials is not complete, its properties will have preferred directions which is not desirable. It is convenient to take the interior node at the centroid. We can also attempt to work only with corner nodes. Since four quantities are necessary for interelement compatibility, one must include derivatives of the function as nodal quantities, i.e.

$$(u)_{\text{node } i} = \left(u \frac{\partial u}{\partial \xi_1} \frac{\partial u}{\partial \xi_2} \right)_{\text{node } i} \tag{3.50}$$

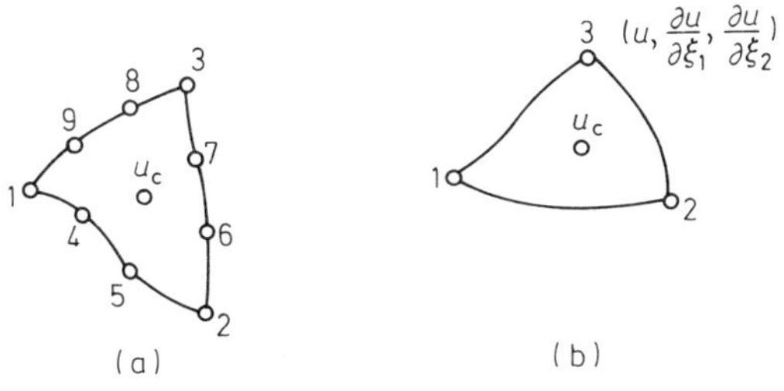

Figure 3.16 Cubic element

As before one additional quantity not associated with the boundaries is required, i.e. the value of u at the centroid, u_c.

3.5 THREE-DIMENSIONAL ELEMENTS

Three-dimensional elements may be useful to represent parts of the body where integration over the domain is required, namely for terms of the type

$$\int_{\Omega} b u^* \, d\Omega \tag{3.51}$$

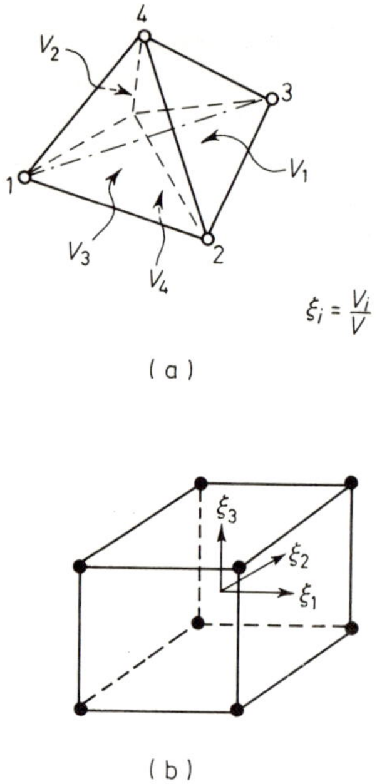

Figure 3.17 First-order three-dimensional elements. (a) *Tetrahedron;* (b) *Cube*

These elements are similar to those used in finite elements and they are well known in the literature. The simplest are the tetrahedron and the cube (Figure 3.17).

TETRAHEDRON

Dimensionless homogeneous coordinates can be defined, similar to those used for the triangular elements seen above. They are (Figure 3.17),

$$\xi_1 = \frac{V_1}{V}, \quad \xi_2 = \frac{V_2}{V}, \quad \xi_3 = \frac{V_3}{V}, \quad \xi_4 = \frac{V_4}{V} \tag{3.52}$$

where the V_i volumes are defined in the figure and V is the volume of the tetrahedron. A function such as b can be written as,

$$b = b_1\xi_1 + b_2\xi_2 + b_3\xi_3 + b_4\xi_4 \tag{3.53}$$

where b_i are the nodal values of b and the interpolation functions are simply ξ_i.

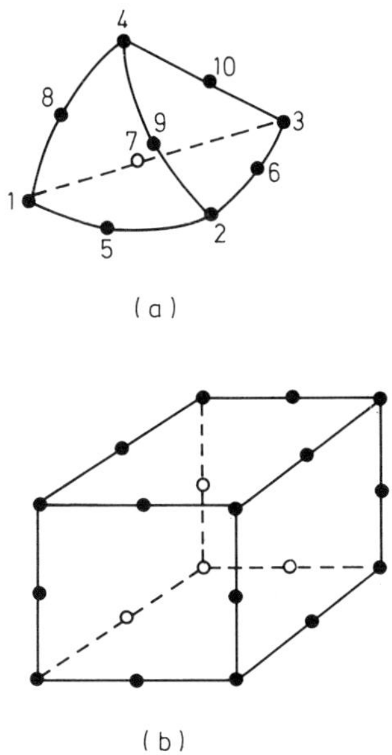

Figure 3.18 Second-order three-dimensional elements. (a) *Tetrahedron;* (b) *Cube*

Tetrahedrons with mid-side nodes can easily be defined (Figure 3.18(a)) and their interpolation functions can be deduced from those of a triangle (equation (3.49)).

CUBE

The simplest cube type element (they are undeformed cubes when plotted in dimensionless coordinates) is shown in Figure 3.17(b). We can also have a quadratic variation of the function on the sides (Figure 3.18(b)).

Interpolation functions for these and other three-dimensional elements can be seen in Connor and Brebbia.[1]

3.6 ORDER OF INTERPOLATION FUNCTIONS

We must always ensure that the expression for u in terms of the curvilinear coordinates contains a complete polynomial in x, y, z. That is by suitably specialising the nodal potentials,

$$u = \sum \phi_i u_i \tag{3.54}$$

we can reproduce

$$u = c_1 + c_2 x + c_3 y + c_4 z \tag{3.55}$$

If the nodal potentials are taken according to (3.55) and substituted into (3.54) we find,

$$u = \sum \phi_i (c_1 + c_2 x_i + c_3 y_i + c_4 z_i) \tag{3.56}$$

Note that in order to be able to reproduce (3.55) everywhere in the element we have to have,

$$\sum \phi_i = 1 \tag{3.57}$$

and

$$\sum x_i \phi_i = x, \qquad \sum y_i \phi_i = y, \qquad \sum z_i \phi_i = z \tag{3.58}$$

Relationship (3.57) is satisfied by any interpolation function, but equations (3.58) are only satisfied if the order of u is at least the same as the order of the x, y, z functions.

The same restriction applies to finite elements as well as boundary ones.

REFERENCES

1. Connor, J. J. and Brebbia, C. A., *Finite Element Techniques for Fluid Flow*, Newnes-Butterworths, London, (1976)
2. Brebbia, C. A. and Wrobel, L., 'The Boundary Element Method'. In *Computer Methods in Fluids*, Ed. C. Taylor, Pentech Press, (1979)

BIBLIOGRAPHY

Bratanow, T., Spehert T. and Brebbia, C. A., 'Three Dimensional Analysis of Fluids around Ship Hulls using Boundary Elements', *Proc. Int. Semin. on Recent Advances in Boundary Elements Methods*, Southampton University, UK, 5–7 July (1978)

4 Fundamental Solutions

4.1 INTRODUCTION

We have seen, in Chapter 2, a few examples of fundamental solutions for a number of simple differential equations. Before we go on to look at more complicated problems it is essential that we first consider their physical significance and study their mathematical properties.

We shall start by introducing the Dirac delta function as a sum of eigenfunctions, and then go on to consider some methods of finding the explicit form of fundamental solutions for some commonly occurring problems.

At the end of the chapter will be found a table of the fundamental solutions for a selection of differential equations (Table 4.1).

4.2 EIGENFUNCTIONS AND THE DIRAC DELTA FUNCTION

Consider a medium in which we have a physical property represented by the function u which may be a function of space and time. If we know the governing equation for u and this equation is linear (non-linear problems will be dealt with in Chapter 6) we can write this equation as

$$\mathscr{L}(u) = 0 \tag{4.1}$$

throughout the medium which we shall for the moment consider to be finite in extent.

Here $\mathscr{L}$ is a linear differential operator, for which we have

$$\mathscr{L}(\alpha\phi_1+\beta\phi_2)=\alpha\mathscr{L}(\phi_1)+\beta\mathscr{L}(\phi_2) \tag{4.2}$$

where ϕ_1 and ϕ_2 are any two functions with sufficient continuity properties to make $\mathscr{L}(\phi_1)$ and $\mathscr{L}(\phi_2)$ meaningful, and α and β are constants.

The eigenfunctions ϕ_n corresponding to $\mathscr{L}$ are defined by

$$\mathscr{L}(\phi_n)=\lambda_n\phi_n \tag{4.3}$$

where λ_n is a constant called the eigenvalue corresponding to the eigenfunction ϕ_n.

The form of the eigenfunctions will depend on the coordinate system used. For rectangular Cartesian coordinates we can choose the eigenfunctions to be the harmonic functions cos (nx), sin (nx), cos (ny), etc., or using complex notation $e^{\pm inx}$. The eigenvalues λ_n are then just polynomials involving powers of 'n' and 'in' for an equation with constant coefficients.[1]

It is usual to choose our eigenfunctions to be a complete set, in the sense that for any function u a number N can be found such that

$$\left\|u-\sum_{1}^{N}\alpha_n\phi_n\right\|<\varepsilon \tag{4.4}$$

where ε is any small number and α_n are constants.

Here we have used '$\| \quad \|$' to denote the norm of a function defined by

$$\|u\|=\sqrt{\int_\Omega u\hat{u}\,\mathrm{d}\Omega} \tag{4.5}$$

where the integration is over all the variables in u on the problem region, Ω and ^ denotes complex conjugation. Equation (4.4) means that we can expand any sufficiently continuous function u in terms of a series of eigenfunctions to any desired degree of accuracy, i.e.

$$u=\lim_{N\to\infty}\left(\sum_{n=0}^{N}\alpha_n\phi_n\right) \tag{4.6}$$

In this chapter we shall use the notation

$$\langle u,v\rangle_\Omega=\int_\Omega u\hat{v}\,\mathrm{d}\Omega \tag{4.7}$$

$$\langle u,v\rangle_\Gamma=\int_\Gamma u\hat{v}\,\mathrm{d}\Gamma \tag{4.8}$$

to represent integration of the two functions u and v over the region Ω and boundary Γ respectively. The expression $\langle u, v \rangle_\Omega$ which will be a number, is referred to as the 'scalar product' of u and v.

The norm of a function u may now be written

$$||u|| = \sqrt{\langle u, u \rangle_\Omega} \tag{4.9}$$

Given a complete set of eigenfunctions we may form an orthonormal set which has the property that

$$\langle \phi_m, \phi_n \rangle_\Omega = \tilde{\delta}_{mn} \tag{4.10}$$

where $\tilde{\delta}_{mn}$ is the Krönecker delta defined by

$$\tilde{\delta}_{mn} = \begin{cases} 0 & \text{if } m \neq n \\ 1 & \text{if } m = n \end{cases} \tag{4.11}$$

We may now define the Dirac delta function on the region in terms of our complete orthonormal set for this region, by

$$\delta_\Omega(\vec{x} - \vec{\xi}) = \sum_{n=0}^{\infty} \phi_n(\vec{x}) \hat{\phi}_n(\vec{\xi}) \tag{4.12}$$

where $\vec{x}$ and $\vec{\xi}$ are two points in space. Hereafter we shall refer to $\vec{\xi}$ as the 'source' point and $\vec{x}$ as the observation point.

For the case

$$\phi_n = \frac{1}{2a} \exp\left(in \frac{\pi x}{a} \right)$$

(the eigenfunctions corresponding to the one-dimensional Laplace equation over the interval $-a$ to a) the delta function for this region will be represented by:

$$\begin{aligned} \delta_\Omega(x - \xi) &= \frac{1}{4a^2} \sum_{n=-\infty}^{\infty} \exp\left(\frac{in\pi x}{a} \right) \exp\left(-\frac{in\pi \xi}{a} \right) \\ &= \frac{1}{4a^2} \sum_{n=-\infty}^{\infty} \exp\left(\frac{in\pi}{a} (x - \xi) \right) \end{aligned} \tag{4.13}$$

By the nature of the expansion we have assumed that this function is periodic with period $2a$, so if we plot this function in the range $-a$ to a we will have ignored 'image' functions repeating outside the region at distances of $2na$.

In Figures 4.1(a), (b) and (c) we have plotted the delta function for $a = \pi$ truncating at 5, 10 and 20 terms respectively. Without loss of generality we have taken our source point ξ to be at the origin. We see that the more terms we take the higher and narrower the peak becomes. If we take the complete series, our peak becomes infinitely narrow and infinitely high, we obtain a generalised function which is $1/2a$ except at the source point where it is infinite.

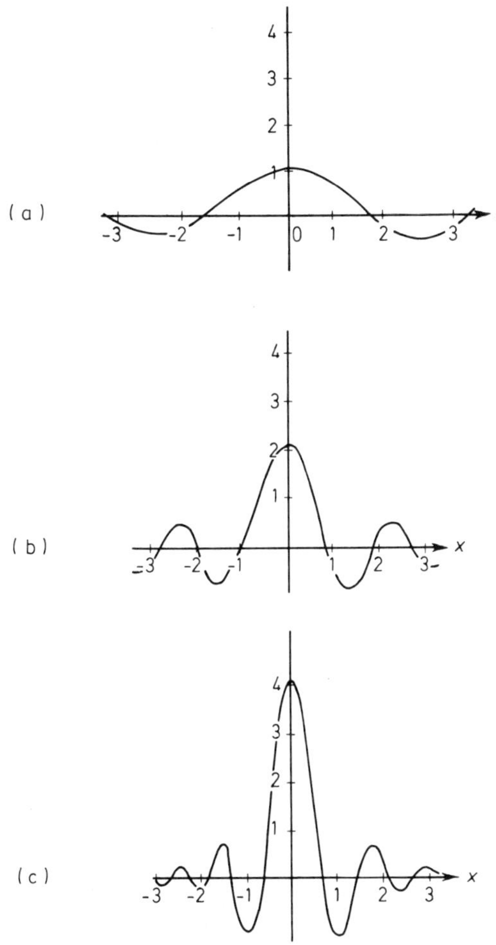

Figure 4.1 The delta function over a finite interval truncated at: (a) *5;* (b) *10; and* (c) *20 terms*

We can see that as a 'forcing function' on the right-hand side of an equation this would give for the solution the field due to a series of sources separated by twice our normalisation range.

We shall illustrate the definitions given above with a simple example.

Example 4.1

Consider the Laplace equation defined on the rectangular region shown in Figure 4.2.

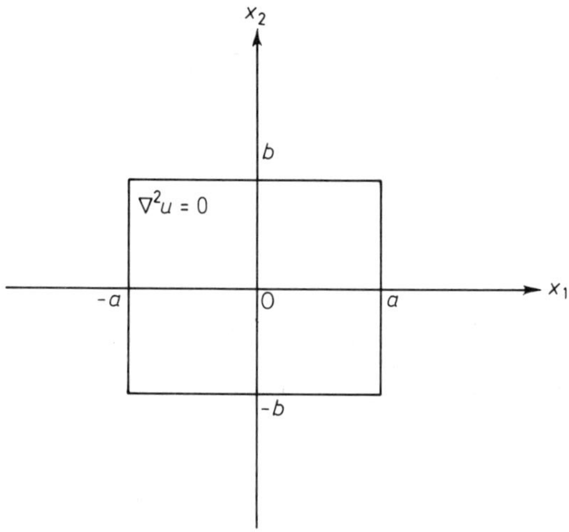

Figure 4.2 Problem region for Example 4.1

We have,

$$\frac{\partial^2 u}{\partial x_1^2}+\frac{\partial^2 u}{\partial x_2^2}=0 \qquad (-a<x_1<a) \quad (-b<x_2<b) \tag{a}$$

The eigenfunctions for this region may be written in complex notation

$$\phi_n(x_1, x_2)=\frac{1}{2\sqrt{ab}}\exp\left(\frac{i\pi n x_1}{a}\right)\exp\left(\frac{i\pi n x_2}{b}\right) \tag{b}$$

as

$$\frac{\partial^2 \phi_n}{\partial x_1^2} + \frac{\partial^2 \phi_n}{\partial x_2^2} = -\frac{\pi^2 n^2}{2\sqrt{ab}} \left(\frac{1}{a^2} + \frac{1}{b^2} \right) \phi_n \tag{c}$$

The nth eigenvalue is thus given by:

$$\lambda_n = -\frac{\pi^2 n^2}{2\sqrt{ab}} \left(\frac{1}{a^2} + \frac{1}{b^2} \right) \tag{d}$$

These eigenfunctions are orthonormal as

$$\begin{aligned}
\langle \phi_m, \phi_n \rangle_\Omega &= \int_{-a}^{a} \int_{-b}^{b} \phi_m \hat{\phi}_n \, \mathrm{d}x_2 \, \mathrm{d}x_1 \\
&= \frac{1}{4ab} \int_{-a}^{a} \int_{-b}^{b} \exp\left(\frac{+\mathrm{i}\pi n x_1}{a} \right) \exp\left(\frac{+\mathrm{i}\pi n x_2}{b} \right) \times \\
&\quad \times \exp\left(\frac{-\mathrm{i}\pi m x_1}{a} \right) \exp\left(\frac{-\mathrm{i}\pi m x_2}{b} \right) \mathrm{d}x_2 \, \mathrm{d}x_1 \\
&= \frac{1}{4ab} \int_{-a}^{a} \exp\left(-\mathrm{i}\pi (m-n) \frac{x_1}{a} \right) \mathrm{d}x_1 \times \\
&\quad \times \int_{-b}^{b} \exp\left(-\mathrm{i}\pi (m-n) \frac{x_2}{b} \right) \mathrm{d}x_2 \\
&= \left\{ \begin{matrix} 1 & \text{for } m = n \\ 0 & \text{for } m \neq n \end{matrix} \right\} = \tilde{\delta}_{mn}
\end{aligned} \tag{e}$$

The eigenfunctions form a complete set as the expansion in (4.6) becomes:

$$u(x_1, x_2) = \frac{1}{2\sqrt{ab}} \sum_{n=-\infty}^{n=+\infty} \sum_{m=-\infty}^{m=+\infty} \alpha_{nm} \exp\left(\frac{\mathrm{i}\pi n x_1}{a} \right) \exp\left(\frac{\mathrm{i}\pi m x_2}{b} \right) \tag{f}$$

which is the complex form of the double Fourier expansion of the function u over the rectangular region.

Note. The expansion (f) assumes that u is a periodic function of x_1 and x_2 with periods $2a$ and $2b$ respectively. This does not concern us here as we are only interested in solving the equation inside the rectangle.

In this region we can write our Dirac delta function:

$$\delta_\Omega(\vec{x}-\vec{\xi}) = \delta_\Omega(x_1-\xi_1)\delta_\Omega(x_2-\xi_2)$$

$$= \frac{1}{4ab}\sum_{n=-\infty}^{\infty} \exp\left(+\frac{\mathrm{i}\pi n}{a}(x_1-\xi_1)\right)\times$$

$$\times \exp\left(+\frac{\mathrm{i}\pi n}{b}(x_2-\xi_2)\right) \qquad \text{(g)}$$

where our source point is (ξ_1, ξ_2) and our observation point is (x_1, x_2).

Note. Again we have a periodic function of x and y with periods $2a$ and $2b$ representing point sources at intervals of $2a$ in the x direction and $2b$ in the direction of y (see Figure 4.3).

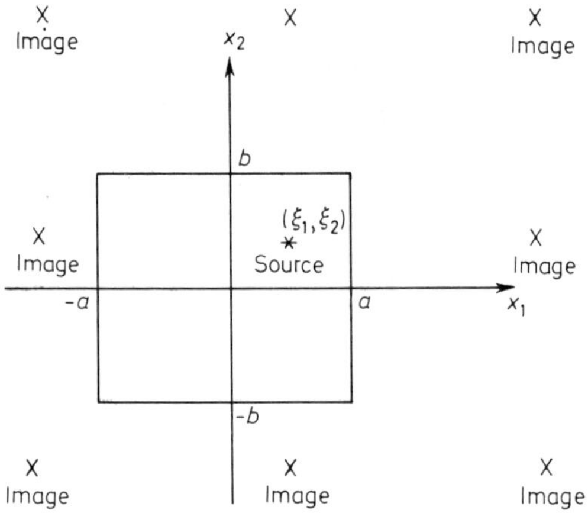

Figure 4.3 Source and 'images' of delta function in region

Consider now an expansion of u in our region Ω in terms of our eigenfunctions ϕ_n given by:

$$u(\vec{x}) = \sum_{n=-\infty}^{\infty} \alpha_n \phi_n \qquad (4.14)$$

In two dimensions we shall interpret the summation $\sum_{n=-\infty}^{\infty} \alpha_n \phi_n$ to mean the double summation $\sum_{n_1=-\infty}^{\infty} \sum_{n_2=-\infty}^{\infty} \alpha_{n_1 n_2} \phi_{n_1 n_2}$ where $\vec{n} = (n_1, n_2)$. For simplicity we shall omit the vector sign off the suffices.

Now consider

$$\langle u(x_1, x_2), \delta_\Omega(\vec{x} - \vec{\xi}) \rangle_{\Omega(\vec{x})} \tag{d}$$

for our case this becomes

$$\int_{-a}^{a} \int_{-b}^{b} u(x_1, x_2) \hat{\delta}_\Omega(\vec{x} - \vec{\xi}) \, dx_1 \, dx_2$$

$$= \frac{1}{4ab} \sum_{n=0}^{\infty} \int_{-a}^{a} \int_{-b}^{b} u(x_1, x_2) \exp\left(-\frac{i\pi n}{a}(x_1 - \xi_1)\right) \times$$

$$\times \exp\left(-\frac{i\pi n}{b}(x_2 - \xi_2)\right) dx_1 \, dx_2$$

$$= \frac{1}{2\sqrt{ab}} \sum_{n=0}^{\infty} \left[\exp\left(+\frac{i\pi n \xi_1}{a}\right) \exp\left(+\frac{i\pi n \xi_2}{b}\right) \frac{1}{2\sqrt{ab}} \times \right.$$

$$\left. \times \int_{-a}^{a} \int_{-b}^{b} u(x_1, x_2) \exp\left(-\frac{i\pi n x_1}{a}\right) \exp\left(-\frac{i\pi n x_2}{b}\right) dx_1 \, dx_2 \right] \tag{e}$$

The integral appearing here is again the form of the Fourier coefficient α_n in the expansion of u, so we may write (e) as

$$\langle u(x_1, x_2), \delta_\Omega(\vec{x} - \vec{\xi}) \rangle_{\Omega(\vec{x})} = \frac{1}{2\sqrt{ab}} \sum_{n=0}^{\infty} \exp\left(+\frac{i\pi n \xi_1}{a}\right) \times$$

$$\times \exp\left(+\frac{i\pi n \xi_2}{b}\right) \alpha_n \tag{f}$$

but this is just the Fourier expansion of $u(\xi_1, \xi_2)$ hence we have demonstrated a special case of equation (4.18), i.e.

$$\langle u(x_1, x_2), \delta_\Omega(\vec{x} - \vec{\xi}) \rangle_{\Omega(\vec{x})} = u(\xi_1, \xi_2) \tag{g}$$

Multiplying by $\hat{\phi}_m$ and integrating over Ω we obtain:

$$\langle u,\phi_m \rangle_\Omega = \left\langle \sum_{n=-\infty}^{\infty} \alpha_n \phi_n, \phi_m \right\rangle_\Omega$$

$$= \sum_{n=-\infty}^{\infty} \alpha_n \langle \phi_n, \phi_m \rangle_\Omega$$

$$= \sum_{n=-\infty}^{\infty} \alpha_n \tilde{\delta}_{nm} = \alpha_m \tag{4.15}$$

$$\alpha_m = \langle u_1, \phi_m \rangle_\Omega \tag{4.16}$$

an expression for our expansion coefficients α_m.

Note. For the case $\phi_m = (2\pi)^{-1} e^{i\pi m x}$ (the eigenfunctions for the one-dimensional Laplace equation, over the interval $-\pi$ to π) the left-hand side corresponds to a finite Fourier transform of u giving the Fourier coefficient α_m.

Now consider the expression

$$\langle u(\vec{x}), \delta_\Omega(\vec{x}-\vec{\xi}) \rangle_\Omega \tag{4.17}$$

we may write both terms as expansions in terms of our ϕ_n:

$$\left\langle \sum_{n=0}^{\infty} \alpha_n \phi_n(\vec{x}), \sum_{m=0}^{\infty} \phi_m(\vec{x}) \hat{\phi}_m(\vec{\xi}) \right\rangle_{\Omega(\vec{x})}$$

$$= \sum_{n=0}^{\infty} \alpha_n \sum_{m=0}^{\infty} \phi_m(\vec{\xi}) \langle \phi_n(\vec{x}), \phi_m(\vec{x}) \rangle_{\Omega(\vec{x})}$$

$$= \sum_{n=0}^{\infty} \alpha_n \sum_{m=0}^{\infty} \phi_m(\vec{\xi}) \tilde{\delta}_{nm}$$

$$\sum_{n=0}^{\infty} \alpha_n \phi_n(\vec{\xi}) = u(\vec{\xi})$$

Hence

$$\langle u(\vec{x}), \delta_\Omega(\vec{x}-\vec{\xi}) \rangle_{\Omega(\vec{x})} = u(\vec{\xi}) \tag{4.18}$$

We have now derived the selective property of a delta function which we shall use extensively in what follows.

Note. For real functions the hat ^ denoting complex conjugation becomes superfluous.

Example 4.2

Considering again our function on the rectangle $-a \leqslant x_1 \leqslant a$, $-b \leqslant x_2 \leqslant b$, we have

$$\delta_\Omega(\vec{x}-\vec{\xi}) = \frac{1}{4ab}\sum_{n=0}^{\infty} \exp\left(\frac{i\pi n}{a}(x_1-\xi_1)\right)\exp\left(\frac{i\pi n}{b}(x_2-\xi_2)\right) \quad \text{(a)}$$

Let $u(x, y)$ be any function on the rectangle. We may write

$$u(x_1, x_2) = \frac{1}{2\sqrt{ab}}\sum_{n=0}^{\infty} \alpha_n \exp\left(\frac{i\pi n x_1}{a}\right)\exp\left(\frac{i\pi n x_2}{b}\right) \quad \text{(b)}$$

where α_n are complex constants given by

$$\alpha_n = \langle u, \phi_n \rangle_\Omega$$

$$= \frac{1}{2\sqrt{ab}}\int_{-a}^{a}\int_{-b}^{b} u(x_1, x_2)\exp\left(-\frac{i\pi n x_1}{a}\right)\exp\left(-\frac{i\pi n x_2}{b}\right)\mathrm{d}x_1\mathrm{d}x_2$$

$$\text{(c)}$$

which in this case is the form of the coefficients in a double Fourier series in terms of a double 'Fourier transform' of u.

4.3 FUNDAMENTAL SOLUTIONS ON A FINITE REGION

We are now in a position to define the fundamental solution $u^*_\Omega(\vec{x}, \vec{\xi})$ for a linear operator $\mathscr{L}$ in a region Ω by

$$\mathscr{L}(u^*_\Omega(\vec{x}, \vec{\xi})) = \delta_\Omega(\vec{x}-\vec{\xi}) \quad (4.19)$$

$\mathscr{L}$ involves differentials in the variables $\vec{x}$. If we expand u^*_Ω and δ_Ω in terms of eigenfunctions: putting

$$u^*_\Omega(\vec{x}, \vec{\xi}) = \sum_{n=0}^{\infty} \alpha_n(\vec{\xi})\,\phi_n(\vec{x})$$

we obtain

$$\mathscr{L}\left(\sum_{n=0}^{\infty} \alpha_n(\vec{\xi})\,\phi_n(\vec{x})\right) = \sum_{n=0}^{\infty} \phi_n(\vec{x})\,\hat{\phi}_n(\vec{\xi}) \tag{4.20}$$

with

$$\sum_{n=0}^{\infty} \alpha_n(\vec{\xi})\,\mathscr{L}(\phi_n) = \sum_{n=0}^{\infty} \alpha_n(\vec{\xi})\,\lambda_n\phi_n(\vec{x})$$

hence

$$\sum_{n=0}^{\infty} \alpha_n(\vec{\xi})\,\lambda_n\phi_n(\vec{x}) = \sum_{n=0}^{\infty} \phi_n(\vec{x})\,\hat{\phi}_n(\vec{\xi}) \tag{4.21}$$

This is an identity true for all $\vec{x}$ and $\vec{\xi}$. Hence we may deduce

$$\alpha_n(\vec{\xi}) = \frac{\hat{\phi}_n(\vec{\xi})}{\lambda_n} \tag{4.22}$$

This formula may also be deduced by using the orthogonality of our eigenfunctions. Here λ_n is assumed not to be a function of $\vec{x}$, i.e. $\mathscr{L}$ has constant coefficients.

Hence the fundamental solution $u^*_\Omega(\vec{x}, \vec{\xi})$ is given by

$$u^*_\Omega(\vec{x}, \vec{\xi}) = \sum_{n=0}^{\infty} \frac{\hat{\phi}_n(\vec{\xi})}{\lambda_n}\phi_n(\vec{x}) \tag{4.23}$$

This function represents the field due to a point source at $\vec{\xi}$ together with 'image' sources generated by the boundary of the region Ω.

Example 4.3

For our example of Laplace's equation on a rectangle discussed in Example 4.1 and 4.2 we have

$$\lambda_n = -\frac{\pi^2 n^2}{2\sqrt{ab}}\left(\frac{1}{a^2} + \frac{1}{b^2}\right) \tag{a}$$

and

$$\phi_n(\vec{x}) = \frac{1}{2\sqrt{ab}}\exp\left(\frac{\mathrm{i}\pi n x_1}{a}\right)\exp\left(\frac{\mathrm{i}\pi n x_2}{b}\right) \tag{b}$$

hence our fundamental solution is given by

$$u^*_\Omega(\vec{x}, \vec{\xi}) = -\frac{(ab)^{3/2}}{2(a^2+b^2)\pi^2} \sum_{\substack{n=-\infty \\ n \neq 0}}^{\infty} \left[\frac{1}{n^2} \exp\left(\frac{\mathrm{i}\pi n}{a}(x_1-\xi_1)\right) \times \right.$$

$$\left. \times \exp\left(\frac{\mathrm{i}\pi n}{b}(x_2-\xi_2)\right)\right] + \alpha_0 \qquad \text{(c)}$$

We have separated out the term corresponding to $n = 0$ as the $1/n^2$ factor will make this constant term infinite. In practice this is not important as the physically significant variable is the space derivative of *u in potential* problems.

We have seen that the fundamental solution is the solution corresponding to a point source, or in a finite region to a number of sources reflected according to the geometry of the region. We shall see that for two-dimensional Laplace equations the empty space fundamental solution corresponding to an infinite problem region does not decrease to zero with increasing distances from the source, in fact it increases without bounds. In this way 'image' sources outside the problem region will contribute more to the resulting field inside than the internal source itself. This is not the case for more complicated equations for which the Laplace is often an approximation. From the point of view of fundamental solutions in a finite two-dimensional region, the Laplace equation is too crude a representation. To illustrate this point we shall now consider the two-dimensional Helmholtz equation, which reduces to Laplace if we let κ^2 tend to zero.

Example 4.4

Suppose we have the Helmholtz equation in two dimensions on the square $-a \leqslant x_1 \leqslant a$, $-a \leqslant x_2 \leqslant a$:

$$\frac{\partial^2 u}{\partial x_1^2} + \frac{\partial^2 u}{\partial x_2^2} + \kappa^2 u = 0 \qquad \text{(a)}$$

The orthonormal eigenfunctions are

$$\phi_n = \frac{1}{2a} \exp\left(\mathrm{i}\pi n \frac{(x_1+x_2)}{a}\right) \qquad (n = 0, \pm 1, \pm 2, \ldots) \qquad \text{(b)}$$

and the eigenvalues are

$$\lambda_n = \frac{1}{2a}\left(\kappa^2 - \frac{2\pi^2 n^2}{a^2}\right) \tag{c}$$

hence the fundamental solution for this region is

$$u^*_\Omega(\vec{x}, \vec{\xi}) = \frac{1}{4a^2} \sum_{n=-\infty}^{\infty} \frac{1}{\kappa^2 - (2\pi^2 n^2/a^2)} \exp\left(\frac{\mathrm{i}\pi n}{a}(x_1 - \xi_1)\right) \times \\ \times \exp\left(\frac{\mathrm{i}\pi n}{a}(x_2 - \xi_2)\right) \tag{d}$$

Note that now we have no trouble with the constant, $n = 0$, term, we only get zeros in the denominator if $\kappa^2 = 2\pi^2 n^2/a^2$ for some n.

4.4 INFINITE DOMAINS

We have seen that we may expand any function within a domain in terms of a complete set of eigenfunctions orthonormal over the domain by putting

$$u(x_1, x_2) = \sum_{n=-\infty}^{\infty} \alpha_n \phi_n \tag{4.24}$$

The eigenfunctions depended on the coordinate system chosen and the shape of the region.

If we now wish to represent a function over all space we cannot use an expansion similar to (4.24), in this case we have to construct a set of eigenfunctions normalised over all space. This set will now depend on a continuously varying parameter $\vec{\kappa}$, say, rather than just being denoted by a discrete set of subscripts such as n. We shall denote these eigenfunctions by $\phi(\vec{\kappa}, \vec{x})$. These functions will now depend only on the coordinate system used. $\vec{\kappa}$ will be a vector with the same number of components as $\vec{x}$, i.e. in two dimensions the two components κ_1, κ_2 will correspond to the subscripts, n_1 and n_2 in $\phi_{n_1 n_2}$.

We shall see that in infinite space the summations will be replaced by integrations and our discrete indices will be replaced by the now continuously varying parameter $\vec{\kappa}$.

For our infinite region our orthonormalisation condition becomes

$$\int_{\substack{\text{all}\\\text{space}}} \phi(\vec{\kappa}, \vec{x})\hat{\phi}(\vec{\kappa}', \vec{x})\,\mathrm{d}\vec{x} = \langle\, \phi(\vec{\kappa}, \vec{x}),\ \phi(\vec{\kappa}', \vec{x})\,\rangle_{\vec{x}} = \delta(\vec{\kappa} - \vec{\kappa}') \tag{4.25}$$

where the delta function is now the classical Dirac delta function defined for all space, which is zero when $\vec{\kappa} \neq \vec{\kappa}'$ but infinite at $\vec{\kappa} = \vec{\kappa}'$. To define δ uniquely we require that

$$\langle\, \phi(\vec{\kappa}, \vec{x}),\ \delta(\vec{\kappa} - \vec{\kappa}')\,\rangle_{\vec{\kappa}} = \phi(\vec{\kappa}', \vec{x}) \tag{4.26}$$

which is the replacement property of the delta function with respect to our eigenfunctions. One consequence of (4.26) is that

$$\langle\, 1, \delta(\vec{\kappa} - \vec{\kappa}')\,\rangle_{\vec{\kappa}} = 1 \tag{4.27}$$

This function is now expressed in terms of our eigenfunctions by,

$$\delta(\vec{x} - \vec{\xi}) = \langle\, \phi(\vec{\kappa}, \vec{x}),\ \phi(\vec{\kappa}, \vec{\xi})\,\rangle_{\vec{\kappa}} \tag{4.28}$$

The brackets $\langle\ \rangle_{\vec{\kappa}}$ now denote an integration over all $\vec{\kappa}$ space, i.e.

$$\langle \phi(\vec{\kappa}, \vec{x}), \phi(\vec{\kappa}, \vec{\xi})\,\rangle_{\vec{\kappa}} = \int_{\substack{\text{all}\\\text{space}}} \phi(\vec{\kappa}, \vec{x})\,\phi(\vec{\kappa}, \vec{\xi})\,\mathrm{d}\vec{\kappa} \tag{4.29}$$

The expansion of any bounded function will now be given by

$$u(\vec{x}) = \langle\, \alpha(\vec{\kappa}),\ \phi(\vec{\kappa}, \vec{x})\,\rangle_{\vec{\kappa}} \tag{4.30}$$

where by an argument similar to equations (4.13) to (4.16) we have

$$\alpha(\vec{\kappa}) = \langle\, \phi(\vec{\kappa}, \vec{x}),\ u(\vec{x})\,\rangle_{\vec{x}} \tag{4.31}$$

Equations (4.30) and (4.31) are a generalised transform pair.

If we now consider the expression corresponding to (4.17) for the infinite domain

$$\langle\, u(\vec{x}),\ \delta(\vec{x} - \vec{\xi})\,\rangle_{\vec{x}} \tag{4.32}$$

and expand both in terms of eigenfunctions we obtain

$$\langle\, \langle\, \alpha(\vec{\kappa}),\ \phi(\vec{\kappa}, \vec{x})\,\rangle_{\vec{\kappa}},\ \langle\, \phi(\vec{\kappa}', \vec{x}),\ \phi(\vec{\kappa}', \vec{\xi})\,\rangle_{\vec{\kappa}}\,\rangle_{\vec{x}} \tag{4.33}$$

assuming convergence of the integrals we may rearrange the above to give

$$
\begin{aligned}
&\langle\langle\langle \phi(\vec{\kappa}, \vec{x}), \phi(\vec{\kappa}', \vec{x}) \rangle_{\vec{x}}, \hat{\alpha}(\vec{\kappa})\phi(\vec{\kappa}, \vec{\xi}) \rangle_{\vec{\kappa}} \rangle_{\vec{\kappa}'} \\
&= \langle\langle \delta(\vec{\kappa} - \vec{\kappa}'), \alpha(\vec{\kappa})\phi(\vec{\kappa}', \vec{\xi}) \rangle_{\vec{\kappa}} \rangle_{\vec{\kappa}'} \\
&\qquad\qquad \text{using (4.26)} \\
&= \langle \alpha(\vec{\kappa}), \langle \delta(\vec{\kappa} - \vec{\kappa}'), \phi(\vec{\kappa}', \vec{\xi}) \rangle_{\vec{\kappa}'} \rangle_{\vec{\kappa}} \\
&= \langle \alpha(\vec{\kappa}), \phi(\vec{\kappa}, \vec{\xi}) \rangle_{\vec{\kappa}} \\
&\qquad\qquad \text{using (4.27)} \\
&= u(\vec{\xi}) \qquad \text{by equation (4.31)}
\end{aligned} \tag{4.34}
$$

Hence

$$
\langle u(\vec{x}), \delta(\vec{x} - \vec{\xi}) \rangle_{\vec{x}} = u(\vec{\xi}) \tag{4.35}
$$

This relation is fundamental in the boundary element method. We shall illustrate the above points with an example.

Example 4.5

In one dimension or in Cartesian coordinates for an n-dimensional space, we may choose $(1/\sqrt{2\pi})\,e^{i\kappa x}$ for our eigenfunctions, where κ is a continuously varying parameter which will be a scalar variable for one dimension. The constant $1/\sqrt{2\pi}$ appears as a result of the 'normalisation' condition given in equation (4.27).

Our delta function may now be written

$$
\delta(x - \xi) = \frac{1}{2\pi}\int_{-\infty}^{\infty} \exp[i\kappa(x - \xi)]\,d\kappa \tag{a}
$$

any bounded function of x can now be written

$$
u(x) = \frac{1}{\sqrt{2\pi}}\int_{-\infty}^{\infty} \tilde{u}(\kappa)e^{-i\kappa x}\,d\kappa \tag{b}
$$

from equation (4.31), where $\tilde{u}(\kappa)$ is now the Fourier transform of u and $\tilde{u}(\kappa)$ is given by

$$
\tilde{u}(\kappa) = \frac{1}{\sqrt{2\pi}}\int_{-\infty}^{\infty} u(x)e^{i\kappa x}\,dx \tag{c}
$$

Equations (b) and (c) now form a Fourier transform pair.

4.5 THE FUNDAMENTAL SOLUTION IN INFINITE SPACE

The fundamental solution in infinite space corresponding to an isolated point source at the point $\vec{\xi}$ is defined by

$$\mathscr{L}(u^*(\vec{x}, \vec{\xi})) = \delta(\vec{x} - \vec{\xi}) \tag{4.36}$$

Expanding u^* and δ in terms of eigenfunctions, i.e. putting

$$u^*(\vec{x}, \vec{\xi}) = \langle \phi(\vec{\kappa}, \vec{x}),\ \hat{\alpha}(\vec{\kappa}, \vec{\xi}) \rangle_{\vec{\kappa}} \tag{4.37}$$

and

$$\delta(\vec{x} - \vec{\xi}) = \langle \phi(\vec{\kappa}, \vec{x}),\ \phi(\vec{\kappa}, \vec{\xi}) \rangle_{\vec{\kappa}} \tag{4.38}$$

we write (4.36)

$$\mathscr{L}(\langle \phi(\vec{\kappa}, \vec{x}),\ \hat{\alpha}(\vec{\kappa}, \vec{\xi}) \rangle_{\vec{\kappa}}) = \langle \phi(\vec{\kappa}, \vec{x}),\ \phi(\vec{\kappa}, \vec{\xi}) \rangle_{\vec{\kappa}} \tag{4.39}$$

$$\therefore \langle \mathscr{L}(\phi(\vec{\kappa}, \vec{x})),\ \hat{\alpha}(\vec{\kappa}, \vec{\xi}) \rangle_{\vec{\kappa}} = \langle \phi(\vec{\kappa}, \vec{x}),\ \phi(\vec{\kappa}, \vec{\xi}) \rangle_{\vec{\kappa}} \tag{4.40}$$

$$\therefore \langle \lambda(\vec{\kappa})\phi(\vec{\kappa}, \vec{x}),\ \hat{\alpha}(\vec{\kappa}, \vec{\xi}) \rangle_{\vec{\kappa}} = \langle \phi(\vec{\kappa}, \vec{x}),\ \phi(\vec{\kappa}, \vec{\xi}) \rangle_{\vec{\kappa}} \tag{4.41}$$

which can be written

$$\langle \phi(\vec{\kappa}, \vec{x}),\ \hat{\lambda}(\vec{\kappa})\hat{\alpha}(\vec{\kappa}, \vec{\xi}) \rangle_{\vec{\kappa}} = \langle \phi(\vec{\kappa}, \vec{x}),\ \phi(\vec{\kappa}, \vec{\xi}) \rangle_{\vec{\kappa}} \tag{4.42}$$

This is an identity for all $\vec{x}$ and $\vec{\xi}$, hence we may write

$$\alpha(\vec{\kappa}, \vec{\xi}) = \frac{\hat{\phi}(\vec{\kappa}, \vec{\xi})}{\lambda(\vec{\kappa})} \tag{4.43}$$

We can compare this directly with equation (4.22) for the discrete case. Hence we can write our fundamental solution

$$u^*(\vec{x}, \vec{\xi}) = \langle \phi(\vec{\kappa}, \vec{x}),\ \phi(\vec{\kappa}, \vec{\xi})/\hat{\lambda}(\vec{\kappa}) \rangle_{\vec{\kappa}} \tag{4.44}$$

This procedure indicates the general method of finding fundamental solutions by transform methods. We may also use the fundamental solution to solve problems like

$$\mathscr{L}(u) = P \tag{4.45}$$

where P is some function of $\vec{x}$ in our region.

If we put

$$u(\vec{x}) = \langle u^*(\vec{x}, \vec{\xi}),\ P(\vec{\xi}) \rangle_{\Omega(\vec{\xi})} \tag{4.46}$$

we obtain equation (4.45) by substituting for u using

$$\mathscr{L}(u^*(\vec{x}, \vec{\xi})) = \delta(\vec{x} - \vec{\xi})$$

Hence (4.46) gives the solution of (4.45).

Furthermore if we have

$$\mathscr{L}(u) = \mathscr{M}(P) \tag{4.47}$$

where $\mathscr{M}$ is some linear differential operator, then

$$\dot{u}(x) = \mathscr{M}(\langle u^*(\vec{x}, \vec{\xi}), P(\vec{\xi}) \rangle_{\Omega}) \tag{4.48}$$

is the solution of (4.47) as can also be verified by substitution. We shall now give a simple example to illustrate these points.

Example 4.6

Consider the one-dimensional reduced wave equation given by

$$\frac{d^2u}{dx^2} + \kappa^2 u = 0 \tag{a}$$

The orthonormal eigenfunctions we choose correspond to our one-dimensional Cartesian system and are given by

$$\phi(\kappa', x) = \frac{1}{\sqrt{2\pi}} e^{-i\kappa' x} \tag{b}$$

and the eigenvalues

$$\lambda(\kappa') = \kappa^2 - \kappa'^2$$

We wish to find the fundamental solution defined by

$$\frac{d^2u^*}{dx^2} + \kappa^2 u^* = \delta(x - \xi) \tag{c}$$

we may write

$$\begin{aligned} \delta(x-\xi) &= \langle \phi(\kappa, x), \phi(\kappa, \xi) \rangle_{\kappa} \\ &= \frac{1}{2\pi} \int_{-\infty}^{\infty} \exp[-i\kappa(x-\xi)]\, d\kappa \end{aligned} \tag{d}$$

and

$$u^*(x,\xi) = \langle \tilde{u}^*(\kappa,\xi), \phi(\kappa, x)\rangle_\kappa$$
$$= \frac{1}{\sqrt{2\pi}} \int_{-\infty}^{\infty} \tilde{u}^*(\kappa,\xi)\, e^{-i\kappa x}\, d\kappa \qquad \text{(e)}$$

$u^*(\kappa, \xi)$ is the Fourier transform of the fundamental solution given by

$$\tilde{u}^*(\kappa,\xi) = \frac{1}{\sqrt{2\pi}} \int_{-\infty}^{\infty} u^*(\kappa,\xi)\, e^{i\kappa x}\, d\kappa \qquad \text{(f)}$$

from (4.32), which is the inverse relationship corresponding to (e).

The fundamental solution is now given by

$$\frac{d^2}{dx^2}\left(\frac{1}{\sqrt{2\pi}} \int_{-\infty}^{\infty} \tilde{u}^*(\kappa',\xi)\, e^{-i\kappa' x}\, d\kappa'\right) + \kappa^2 \frac{1}{\sqrt{2\pi}} \times$$
$$\times \int_{-\infty}^{\infty} \tilde{u}^*(\kappa',\xi)\, e^{-i\kappa' x}\, d\kappa'$$
$$= \frac{1}{2\pi} \int_{-\infty}^{\infty} \exp[i\kappa'(-x+\xi)]\, d\kappa' \qquad \text{(g)}$$

$$\because \int_{-\infty}^{\infty} (\kappa^2 - \kappa'^2)\tilde{u}^*(\kappa',\xi) e^{-i\kappa' x}\, d\kappa'$$
$$= \frac{1}{\sqrt{2\pi}} \int_{-\infty}^{\infty} \exp[i\kappa'(-x+\xi)]\, d\kappa' \qquad \text{(h)}$$

or

$$\int_{-\infty}^{\infty} \left((\kappa^2 - \kappa'^2)\, \tilde{u}^* - \frac{1}{\sqrt{2\pi}} e^{i\kappa'\xi} \right) e^{-i\kappa' x}\, d\kappa' = 0 \qquad \text{(i)}$$

Note that the expression in large parentheses is the equation resulting from taking the Fourier transform of both sides of equation (c).

Note. (i) is an identity for all values of ξ and we may write simply

$$(\kappa^2 - \kappa'^2)\tilde{u}^* = \frac{1}{\sqrt{2\pi}} e^{i\kappa'\xi} \qquad \text{(j)}$$

which gives the Fourier transform of the solution directly.

Alternatively we may exploit the orthonormalisation of our eigenfunctions to retrieve the same result.

We have therefore

$$\tilde{u}^*(\kappa', \xi) = \frac{1}{(\kappa^2 - \kappa'^2)} \frac{1}{\sqrt{2\pi}} e^{i\kappa'\xi} \qquad \kappa \neq \kappa' \tag{k}$$

and hence using (f) we can write

$$u^*(x, \xi) = \frac{1}{2} \int_{-\infty}^{\infty} \frac{1}{(\kappa^2 - \kappa'^2)} \exp\left[i\kappa'(\xi - x)\right] d\kappa' \tag{l}$$

which by simple contour integration gives (taking the real part for our actual solution)

$$u^*(x. \xi) = \frac{1}{2\kappa} \sin(\kappa|x - \xi|) = \frac{1}{2\kappa} \sin(\kappa r) \tag{m}$$

where r is the (positive) distance between x and ξ.

The above example demonstrates the procedure used to obtain the fundamental solution in Cartesian coordinates. The main steps of the analysis are given below:

1 Transform the equation a sufficient number of times in order to obtain an algebraic equation in terms of the transformed function $\tilde{u}^*(\vec{\kappa}, \vec{\xi})$.
2 Solve this equation in terms of $\tilde{u}^*$.
3 Invert the transformations to obtain the fundamental solution $u^*(\vec{x}, \vec{\xi})$.

In practice it may not be necessary to perform the full number of transformations as the resulting transformed equation at some point may have a known fundamental solution. The inverse transforms may then be performed on this function to obtain the fundamental solution for the original equation.

As an example of the non-trivial use of this technique, the fundamental solution for a concentrated load P applied at the point $\vec{\xi}$ in an anisotropic elastic space will now be deduced starting from Navier's equations in the form[3,4]

$$C_{pqrs} u_{r,sq} = 0 \tag{4.49}$$

where C_{pqrs} is the tensor relating stresses to strains and involves the elastic constants of the material, u_r is the displacement in direction x_r and $_{,sq}$ denotes the differentiation $\partial^2/\partial x_s \partial x_q$. The summation convention has been used here, a repeated suffix is assumed to be summed.

It is convenient to define a fundamental tensor U^*_{ji} by

$$u^*_i = P_j U^*_{ji} \tag{4.50}$$

We then require the fundamental solution corresponding to (4.49), i.e.

$$C_{pqrs} U^*_{tr,\,sq} = \delta_t(\vec{x} - \vec{\xi}) \qquad (p = 1, 2, 3) \tag{4.51}$$

where the subscript t indicates that the delta function δ_t is the tth component of some vector delta function, t then represents the direction of the applied force at $\vec{\xi}$. We may take $\vec{\xi}$ to be the origin by a change of axes. Taking the triple Fourier transform of (4.51) we obtain

$$F_{pr}(\vec{\kappa})\, \tilde{U}^*_{rt}(\vec{\kappa}) = \tilde{\delta}_{pt} \tag{4.52}$$

where $\tilde{\delta}_{pt}$ is the Krönecker delta as before.

$$\tilde{U}^*_{rt}(\vec{\kappa}) = \frac{1}{(2\pi)^{3/2}} \int_{-\infty}^{\infty} U_{rt}(\vec{x})\, \mathrm{e}^{\mathrm{i}\vec{\kappa}.\vec{x}}\, \mathrm{d}\vec{x} \tag{4.53}$$

and

$$F_{pr}(\vec{\kappa}) = C_{pqrs} \kappa_q \kappa_s \tag{4.54}$$

Inverting the matrix equation (4.52) to give us the triple Fourier transform of the fundamental tensor we have

$$\tilde{U}^*_{rt}(\vec{\kappa}) = F_{tr}^{-1}(\vec{\kappa}) \tag{4.55}$$

inverting the Fourier transforms we obtain the usual form for the fundamental solution in component form as

$$U^*_{rt}(\vec{x}) = \frac{1}{(2\pi)^{3/2}} \int_{-\infty}^{\infty} (C_{pqrs} \kappa_q \kappa_s)^{-1}\, \mathrm{e}^{-\mathrm{i}\vec{\kappa}.\vec{x}}\, \mathrm{d}\vec{\kappa} \tag{4.56}$$

more usually written

$$U^*_{rt}(\vec{x}) = \frac{1}{(2\pi)^{3/2}} \int_{-\infty}^{\infty} F_{tr}^{-1}\, \mathrm{e}^{-\mathrm{i}\vec{\kappa}.\vec{x}}\, \mathrm{d}\vec{\kappa} \tag{4.57}$$

This can be transformed to a simpler form by taking spherical polar coordinates and integrating twice to give

$$U_{ij}^*(\vec{x}) = \frac{1}{8\pi^2 r} \oint_{|\eta|=1} F_{ij}^{-1}(\eta)\, \mathrm{d}s \tag{4.58}$$

The integral is now taken round a unit circle which has centre at the source point (the origin here) and lying in the plane perpendicular to the line joining $\vec{x}$ to the origin at a distance r.

For a general source point $\vec{\xi}$ (4.58) becomes

$$U_{ij}^*(\vec{x}, \vec{\xi}) = \frac{1}{8\pi^2 |\vec{x} - \vec{\xi}|} \oint_{|\eta|=1} F_{ij}^{-1}(\eta)\, \mathrm{d}s \tag{4.59}$$

For an isotropic material (4.59) may be evaluated to give the Kelvin solution:

$$U_{ij}^*(\vec{x}, \vec{\xi}) = \frac{1}{4\pi\mu|\vec{x} - \vec{\xi}|} \left(\frac{3-4\nu}{4(1-\nu)} \delta_{ij} + \frac{1}{4(1-\nu)} \frac{(x_i - \xi_i)}{|\vec{x} - \vec{\xi}|} \frac{(x_j - \xi_j)}{|\vec{x} - \vec{\xi}|} \right)$$

In the general case (4.59) must be evaluated numerically.

4.6 NUMERICAL SOLUTIONS (ANISOTROPIC BODIES)

We have seen that the determination of the fundamental solution involves taking transforms of both sides of our original equation with respect to the eigenfunctions ϕ_n, and later inverting these transforms.

These processes can only be performed in closed form for a very limited number of special cases. When we have a governing equation with different coefficients such as with anisotropic media, a numerical approach for the evaluation of our solution becomes essential.

Suppose for simplicity we are working in rectangular Cartesian coordinates, then the operator $\mathscr{L}$ may be represented as a polynomial combination of the derivatives $\partial/\partial x_i$, $i = 1, \ldots, n$. We can write

$$\mathscr{L} \equiv \mathrm{P}\left(a_1 \frac{\partial}{\partial x_i}, a_0, a_2 \frac{\partial^2}{\partial x_i \partial x_j}, \ldots \right) \tag{4.61}$$

where a_0, a_1, a_2, ... are constants independent of x. Then the equation for the fundamental solution will be

$$\mathscr{L}(u^*(x, \xi)) = \delta(x - \xi) \tag{4.62}$$

Transforming this equation a sufficient number of times we obtain an equation which for constant a_0, a_1, a_2 is

$$\frac{1}{(2\pi)^{n/2}} P(a_0, \mathrm{i}a_1\kappa_i, -a_2\kappa_i\kappa_j, \ldots)\, \tilde{u}^*(\vec{\kappa}, \vec{\xi}) = 1 \tag{4.63}$$

where n is the number of coordinate directions, hence

$$\tilde{u}^*(\vec{\kappa}, \vec{\xi}) = \frac{1}{P(a_0, \mathrm{i}a_1\kappa_i, -a_2\kappa_i\kappa_j, \ldots)} \tag{4.64}$$

then the fundamental solution is given by the inverse transform:

$$u^*(\vec{x}, \vec{\xi}) = \frac{1}{(2\pi)^{n/2}} \int_{-\infty}^{\infty} \frac{\tilde{u}^*(\vec{\kappa}, \vec{\xi})\, \mathrm{e}^{-\mathrm{i}\vec{\kappa}, \vec{x}}}{P(a_0, \mathrm{i}a_1\kappa_i, -a_2\kappa_i\kappa_j, \ldots)}\, \mathrm{d}\vec{\kappa} \tag{4.65}$$

In the case where a_0, a_1, . . . are functions of space the transformations must be done numerically, i.e. the integrations involved in the transformations are performed numerically. Equation (4.65) will then consist of tables of values of $u^*(\vec{\kappa}, \vec{\xi})$ for various values of the parameter $\vec{\kappa}$ and the source coordinate $\vec{\xi}$. The inverse transform equation (4.65) then becomes the summation of the integrand for the points where $\tilde{u}^*$ has been calculated.

These numerical integrations will involve a truncation of the range of integration of the integrals which in turn means that an approximation has been made to the Dirac delta function on the right-hand side. Figures 4.4(a) and (b) show the representation of the δ function for various integration ranges.

Example 4.7

Considering the equation of Example 4.6, i.e.

$$\frac{\mathrm{d}^2u}{\mathrm{d}x^2} + \kappa^2 u = 0 \tag{a}$$

we found the fundamental solution defined by

$$\frac{\mathrm{d}^2u}{\mathrm{d}x^2} + \kappa^2 u^* = \delta(x - \xi) \tag{b}$$

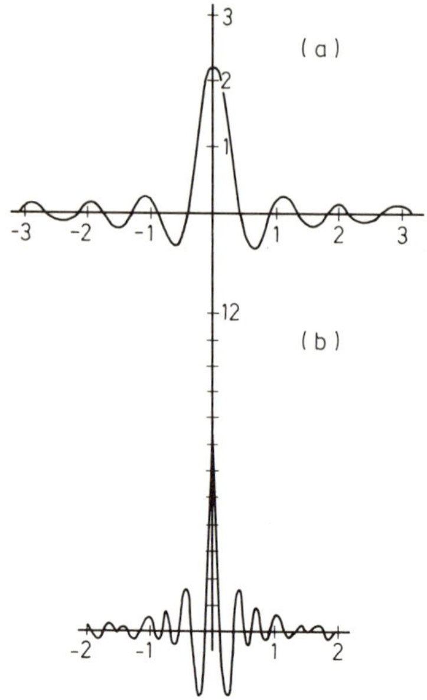

Figure 4.4 Dirac delta function (approximate): (a) *integration range = 7, integration increment = 0.1;* (b) *integration range = 40, integration increment = 0.2*

to be given by

$$u^*(x, \xi) = \frac{1}{2\pi} \int_{-\infty}^{\infty} \frac{1}{(\kappa^2 - \kappa'^2)} \exp[\mathrm{i}\kappa'(\xi - x)] \mathrm{d}\kappa' \qquad \text{(c)}$$

which can be evaluated by contour integration to give

$$u^*(x, \xi) = \frac{1}{2\kappa} \sin(\kappa r)$$

where r is the positive distance between x and ξ. In Figures 4.5(a) and (b) the integral (c) has been represented as a truncated integral which is evaluated numerically. This demonstrates the numerical method of finding u^* which assumes that we know the form of the integrand of equation (c).

4.7 THE METHOD OF IMAGES

So far we have a method of finding the fundamental solution in empty space. In some circumstances our problem region may be confined in

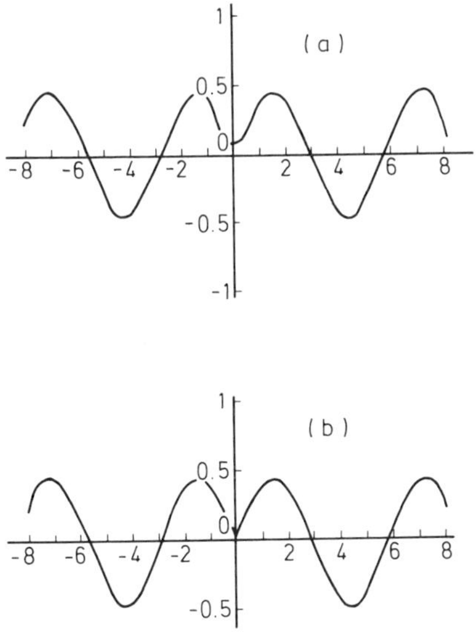

Figure 4.5 Approximate fundamental solution for the one-dimensional Helmholtz equation, $\kappa = 1.1$*:* (a) *integration range = 5, integration increment = 0.2;* (b) *integration range = 20, integration increment = 0.2*

some regular way and it may be more convenient to find a fundamental solution specific to a region.

The simplest case is that of a semi-infinite space such as may occur in a foundation or fluids problem. The surface of the soil or fluid may make it more convenient to work with a semi-infinite space fundamental solution. We choose this solution to satisfy the boundary condition on the interface identically; in this way we shall not need to put elements on the surface when using the boundary integral method.

To obtain these solutions it is necessary to consider their physical interpretation. Consider for the moment a two-dimensional space and take coordinates as in Figure 4.6. The interface is along the x_1 axis and the normal to this is measured in the positive x_2 direction. We shall require that $\partial u/\partial n = 0$ on this interface.

The fundamental solution is the field due to a point source in infinite space and so the field in the presence of our boundary $x_2 = 0$ will represent our half space fundamental solution. Consider a source strength m at a point $(\xi_1, -a)$.

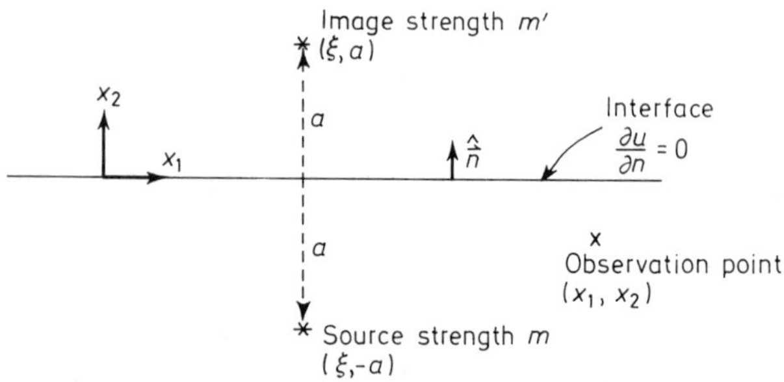

Figure 4.6 Derivation of the fundamental solution in a half plane

The field mu^* from this source will be reflected somehow at $x_2 = 0$, depending on the boundary condition applied there. In order to represent this reflection we place an image source of strength m' behind the interface. The strength of m' required will depend on the kind of interface to be represented. We may then forget the existence of the surface ($x_2 = 0$) and adjust m' in order to satisfy the boundary condition there. The observed field at any point (x_1, x_2) will then be the superposition of the field from both sources (as we have assumed a linear operator for our governing differential equation).

This field will be given by

$$u(x_1) = m\, u^*(x_1, x_2;\, \xi_1, -a) + m'\, u^*(x_1, x_2; \xi_1, a) \tag{4.66}$$

where $u^*(x_1, x_2; \xi_1, \xi_2)$ is the empty space fundamental solution for a source at (ξ_1, ξ_2) observed at (x_1, x_2). Then applying the boundary condition

$$\frac{\partial u}{\partial n} = \frac{\partial u}{\partial x_2} = 0 \qquad \text{on } x_2 = 0 \tag{4.67}$$

we obtain

$$m \frac{\partial u^*(x_1, 0;\, \xi_1, -a)}{\partial x_2} + m' \frac{\partial u^*(x_1, 0;\, \xi_1, a)}{\partial x_2} = 0 \tag{4.68}$$

which for the two-dimensional Laplace equation becomes

$$m \frac{\partial}{\partial x_2}[\ln \sqrt{x_1^2 + (x_2 - a)^2}\,] + m' \frac{\partial}{\partial x_2}[\ln \sqrt{x_1^2 + (x_2 + a)^2}\,]$$
$$= 0 \qquad \text{on } x_2 = 0,\ \forall\, x_1 \tag{4.69}$$

or in particular at $x_1 = 0$ gives

$$\frac{m}{-a} + \frac{m'}{a} = 0 \tag{4.70}$$

Hence we choose $m' = m$. The field is then

$$u = \frac{m}{2\pi}\left\{\ln\left[\sqrt{x_1^2 + (x_2 - a)^2}\right] + \ln\left[\sqrt{x_1^2 + (x_2 + a)^2}\right]\right\} \tag{4.71}$$

$$= \frac{m}{2\pi}\ln\left\{\sqrt{[x_1^2 + (x_2 - a)^2][x_1^2 + (x_2 + a)^2]}\right\} \tag{4.72}$$

Thus the fundamental solution for the two-dimensional Laplace equation for the lower half plane is (Figure 4.6)

$$u_s^*(x_1, x_2; \xi_1, \xi_2) = \frac{1}{2\pi}\ln\left\{\sqrt{[(x_1 - \xi_1)^2 + (x_2 - \xi_2)^2]} \times \right.$$
$$\left. \times \sqrt{[(x_1 - \xi_1)^2 + (x_2 + \xi_2)^2]}\right\} \tag{4.73}$$

Note. We have that u_s^* is the solution of

$$u_s^* = m\delta(x_1, \underset{\substack{\uparrow \\ \text{source}}}{x_2 - a}) + m\delta(x_1, \underset{\substack{\uparrow \\ \text{image}}}{x_2 + a}) \tag{4.74}$$

We could equally have solved this equation in the usual way.

For different boundary conditions we obtain a slightly different equation for m, i.e. if $u = 0$ on $x_2 = 0$ then $m = m'$ and

$$u_s^* = \frac{1}{2\pi}\ln\left[\left(\frac{x_1^2 + (x_2 - \xi_2)^2}{x_1^2 + (x_2 + \xi_2)^2}\right)^{1/2}\right] \tag{4.75}$$

in this case.

The other two important cases are shown in Figures 4.7 and 4.8. In Figure 4.7 we have a parallel layered medium such as might occur in a soils problem. Image sources are constructed as shown. For non-parallel layers a numerical treatment might be appropriate.

Figure 4.8 shows a spherical or cylindrical cavity or obstruction such as might occur in diffraction or fracture problem. In this case an image is constructed at the inverse point given by

$$\mathrm{OI} \times \mathrm{OS} = a^2 \tag{4.76}$$

where a is the radius of the sphere or cylinder.

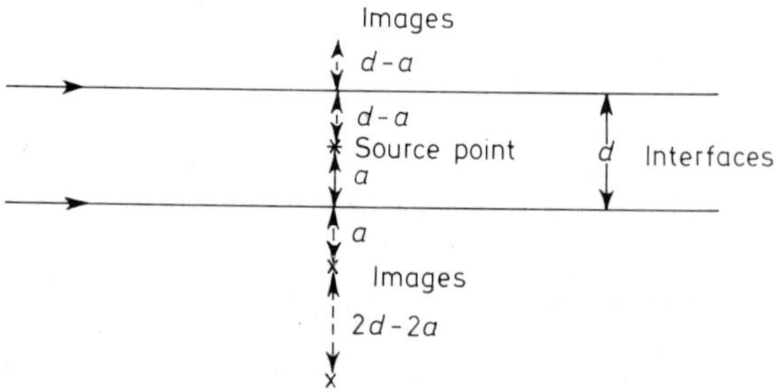

Figure 4.7 Derivation of the fundamental solution in a layer

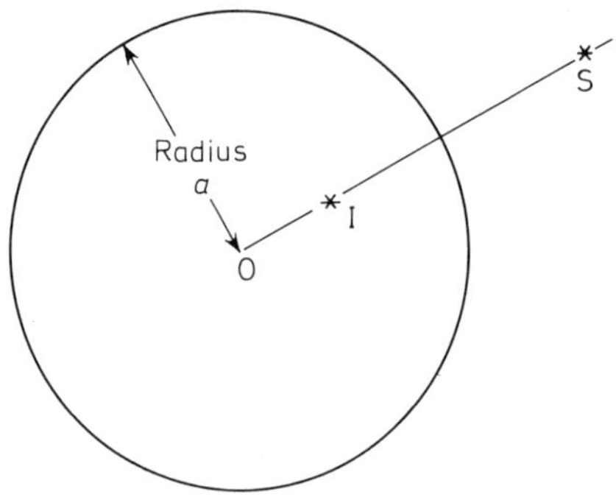

Figure 4.8 Spherical or cylindrical geometry: I, image point; S, source

The most widely used application of this method so far has been for semi-infinite elastic space problems where the Mindlin solution is appropriate.[3]

We shall now deduce this fundamental solution from the Kelvin solution using the method of images. Mindlin considered the problem of finding the field (fundamental solution) due to point forces perpendicular and parallel to the surface of a semi-infinite elastic half space. The boundary condition on this surface was that there was no stress, this gives rise to a system of 'images' of the Kelvin type which in an infinite elastic space would have the effect of creating a situation of zero stress at the position of the interface.

Here we shall consider the case of a point force P normal to the interface at a depth c (see Figure 4.9).

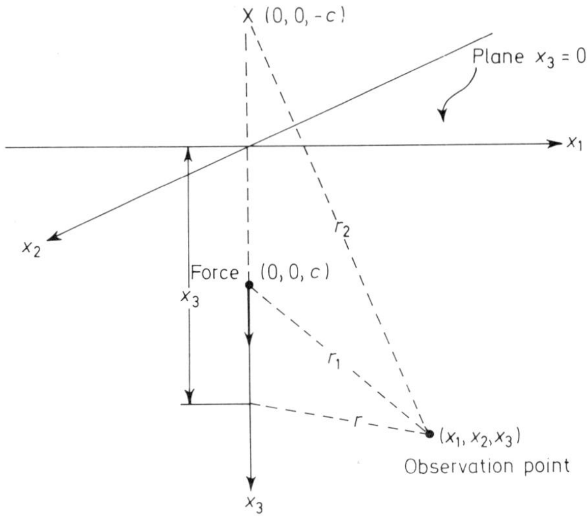

Figure 4.9 Force normal to the boundary in the interior of a semi-infinite solid

Mindlin[3] found that the zero stress condition on $x_3 = 0$ is produced by a combination of images at $(0, 0, -c)$ of the following types:

(a) a force in the x_3 direction at $(0, 0, -c)$
+ (b) a double force in the x_3 direction at $(0, 0, -c)$
+ (c) a centre of compression (three mutually perpendicular double forces) at $(0, 0, -c)$
+ (d) a line of compression along the z axis from $x_3 = -c$ to $x_3 = +c$.
+ (e) a doublet at $(0, 0, -c)$.

Superposing the displacement fields due to the original force and its images we obtain the displacement field for the force normal to the interface for the horizontal directions x_1 and x_2 as

$$U_1^* = U_2^* = \frac{Pr}{16\pi G(1-\mu)}\left(\frac{x_3 - c}{r_1^3} + \frac{(3-4\mu)(x_3-c)}{r_2^3} - \right.$$

$$\left. - \frac{4(1-\mu)(1-2\mu)}{r_2(r_2+x_3+c)} + \frac{6cx_3(x_3+c)}{r_2^3}\right) \tag{4.77}$$

The x_3 component of displacement U_3 is given by:

$$U_3^* = \frac{P}{16\pi G(1-\mu)}\left(\frac{3-4\mu}{r_1} + \frac{8(1-\mu)^2-(3-4\mu)}{r_2} + \frac{(x_3-c)^3}{r_1^3} + \right.$$

$$\left. + \frac{(3-4\mu)(x_3+c)^2 - 2cx_3}{r_2^3} + \frac{6cx_3(x_3+c)^2}{r_2^5}\right) \tag{4.78}$$

These are the Mindlin fundamental solutions for the displacement corresponding to a concentrated force in the positive x_3 direction.

The corresponding stresses using cylindrical polar coordinates (r_1, θ_1, x_3) corresponding to our Cartesian axes (x_1, x_2, x_3) are given by

$$\sigma_{rr}^* = \frac{P}{8\pi(1-\mu)}\left(\frac{(1-2\mu)(x_3-c)}{r_1^3} - \frac{(1-2\mu)(x_3+7c)}{r_2^3} + \right.$$

$$+ \frac{4(1-\mu)(1-2\mu)}{r_2(r_2+x_3+c)} - \frac{3r^2(x_3-c)}{r_1^5} +$$

$$+ \frac{6c(1-2\mu)(x_3+c)^2 - 6c^2(x_3+c) - 3(3-4\mu)r^2(x_3-c)}{r_2^2} -$$

$$\left. - \frac{30cr^2x_3(x_3+c)}{r_2^7}\right) \tag{4.79}$$

$$\sigma_{\theta\theta}^* = \frac{P(1-2\mu)}{8\pi(1-\mu)}\left(\frac{x_3-c}{r_1^3} + \frac{(3-4\mu)(z+c)-6c}{r_2^3} - \frac{4(1-\mu)}{r_2(r_2+x_3+c)} + \right.$$

$$\left. + \frac{6c(x_3+c)^2}{r_2^5} - \frac{6c^2(x_3+c)}{(1-2\mu)r_2^5}\right) \tag{4.80}$$

$$\sigma^*_{zz} = \frac{P}{8\pi(1-\mu)}\left(-\frac{(1-2\mu)(x_3-c)}{r_1^3} + \frac{(1-2\mu)(x_3-c)}{r_2^3} - \frac{3(x_3-c)^3}{r_1^5} - \right.$$

$$-\frac{3(3-4\mu)x_3(x_3+c)^2 - 3c(x_3+c)(5x_3-c)}{r_2^5} -$$

$$\left. -\frac{30cx_3(x_3+c)^3}{r_2^7}\right) \tag{4.81}$$

$$\sigma^*_{rz} = \frac{Pr}{8\pi(1-\mu)}\left(-\frac{1-2\mu}{r_1^3} + \frac{1-2\mu}{r_2^3} - \frac{3(x_3-c)^2}{r_1^5} - \right.$$

$$\left. -\frac{3(3-4\mu)x_3(x_3+c) - 3c(3x_3+c)}{r_2^5} - \frac{30cx_3(x_3+c)}{r_2^7}\right) \tag{4.82}$$

The corresponding formulae for the situation of a concentrated force parallel to the interface at point $x_3 = +c$ are given by:

$$U_1^* = \frac{P}{16\pi G(1-\mu)}\left[\frac{3-4\mu}{r_1} + \frac{1}{r_2} + \frac{x_1^2}{r_1^3} + \frac{(3-4\mu)x_1^2}{r_2^3} + \frac{2cx_3}{r_2^3}\left(1 - \frac{3x_1^2}{r_2^2}\right) + \right.$$

$$\left. + \frac{4(1-\mu)(1-2\mu)}{r_2+x_3+c}\left(1 - \frac{x_1^2}{r_2(r_2+x_3+c)}\right)\right] \tag{4.83}$$

$$U_2^* = \frac{Px_1x_2}{16\pi G(1-\mu)}\left(\frac{1}{r_1^3} + \frac{3-4\mu}{r_2^3} - \frac{6cx_3}{r_2^5} - \frac{4(1-\mu)(1-2\mu)}{r_2(r_2+x_3+c)^2}\right) \tag{4.84}$$

$$U_3^* = \frac{Px_1}{16\pi G(1-\mu)}\left(\frac{x_3-c}{r_1^3} + \frac{(3-4\mu)(x_3-c)}{r_2^3} - \frac{6cx_3(x_3+c)}{r_2^5} + \right.$$

$$\left. + \frac{4(1-\mu)(1-2\mu)}{r_2(r_2+x_3+c)}\right) \tag{4.85}$$

With stresses:

$$\sigma^*_{11} = \frac{Px_1}{8\pi(1-\mu)}\left[-\frac{1-2\mu}{r_1^3} + \frac{(1-2\mu)(5-4\mu)}{r_2^3} - \frac{3x_1^2}{r_1^5} - \frac{3(3-4\mu)x_1^2}{r_2^5} - \right.$$
$$-\frac{4(1-\mu)(1-2\mu)}{r_2(r_2+x_3+c)^2}\left(3 - \frac{x_1^2(3r_2+x_3+c)}{r_2^2(r_2+x_3+c)}\right) +$$
$$\left. + \frac{6c}{r_2^5}\left(3c - (3-2\mu)(x_3+c) + \frac{5x_1^2 x_3}{r_2^2}\right)\right] \tag{4.86}$$

$$\sigma^*_{22} = \frac{Px}{8\pi(1-\mu)}\left[\frac{1-2\mu}{r_1^3} + \frac{(1-2\mu)(3-4\mu)}{r_2^3} - \frac{3x_2^2}{r_1^5} - \frac{3(3-4\mu)x_2^2}{r_2^5} - \right.$$
$$-\frac{4(1-\mu)(1-2\mu)}{r_2(r_2+x_3+c)^2}\left(1 - \frac{x_2^2(3r_2+x_3+c)}{r_2^2(r_2+x_3+c)}\right) +$$
$$\left. + \frac{6c}{r_2^5}\left(c - (1-2\mu)(x_3+c) + \frac{5x_2^2 x_3}{r_2^2}\right)\right] \tag{4.87}$$

$$\sigma^*_{33} = \frac{Px_1}{8\pi(1-\mu)}\left[\frac{1-2\mu}{r_1^3} - \frac{1-2\mu}{r_2^3} - \frac{3(x_3-c)^2}{r_1^5} - \frac{3(3-4\mu)(x_3+c)^2}{r_2^5} + \right.$$
$$\left. + \frac{6c}{r_2^5}\left(c + (1-2\mu)(x_3+c) + \frac{5x_3(x_3+c)^2}{r_2^2}\right)\right] \tag{4.88}$$

$$\sigma^*_{23} = \frac{Px_1x_2}{8\pi(1-\mu)}\left[-\frac{3(x_3-c)}{r_1^5} - \frac{3(3-4\mu)(x_3+c)}{r_2^5} + \right.$$
$$\left. + \frac{6c}{r_2^5}\left(1-2\mu + \frac{5x_3(x_3+c)}{r_2^2}\right)\right] \tag{4.89}$$

$$\sigma^*_{32} = \frac{P}{8\pi(1-\mu)}\left[-\frac{(1-2\mu)(x_3-c)}{r_1^3} + \frac{(1-2\mu)(x_3-c)}{r_2^3} - \right.$$
$$-\frac{3x_1^2(x_3-c)}{r_1^5} - \frac{3(3-4\mu)x_1^2(x_3+c)}{r_2^5} -$$
$$\left. - \frac{6c}{r_2^5}\left(x_3(x_3+c) - (1-2\mu)x_1^2 - \frac{5x_1^2x_3(x_3+c)}{r_2^2}\right)\right] \tag{4.90}$$

$$\sigma_{12}^* = \frac{Px_2}{8\pi(1-\mu)}\left[-\frac{1-2\mu}{r_1^3}+\frac{1-2\mu}{r_2^3}-\frac{3x_1^2}{r_1^5}-\frac{3(3-4\mu)x_1^2}{r_2^5}-\right.$$

$$-\frac{4(1-\mu)(1-2\mu)}{r_2(r_2+x_3+c)^2}\left(1-\frac{x_1^2(3r_2+x_3+c)}{r_2^2(r_2+x_3+c)}\right)-$$

$$\left.-\frac{6cx_3}{r_2^5}\left(1-\frac{5x_1^2}{r_2^2}\right)\right] \quad (4.91)$$

4.8 THE FUNDAMENTAL SOLUTION AND BOUNDARY ELEMENTS

In this section we shall show how the above concepts relate to the boundary element technique.

Consider a problem region Ω, bounded by a surface Γ, in which we have a governing equation

$$\mathscr{L}(u) = 0 \quad (4.92)$$

We wish to find an approximate solution for the function u subject to the boundary conditions

$$\begin{aligned} S(u) &= p \quad \text{on } \Gamma_1 \\ G(u) &= q \quad \text{on } \Gamma_2 \end{aligned} \quad (4.93)$$

$$\Gamma = \Gamma_1 + \Gamma_2$$

where S and G are differential operators.

To solve this problem we may consider the weighted residual form of this problem which can be written

$$\langle \mathscr{L}(u), v \rangle_\Omega = -\langle S(u)-p, \partial v/\partial n \rangle_\Gamma + \langle G(u)-q, v \rangle_\Gamma \quad (4.94)$$

where v is a weighting function, which for the Galerkin formulation is a linear combination of the same basis functions used to build up the approximate solution u. Note that we have not assumed that the boundary conditions (4.78) are identically satisfied.

If we now integrate (4.94) by parts a sufficient number of times to transfer the operator $\mathscr{L}$ to v we obtain

$$\langle u, \hat{\mathscr{L}}(v) \rangle_\Omega = \langle S(u)-p, \hat{G}(v)-q \rangle_\Omega - \langle G(u)-q, \hat{S}(v)-p \rangle_\Gamma \quad (4.95)$$

where $\hat{\mathscr{L}}$, $\hat{G}$ and $\hat{S}$ are the adjoint operators corresponding to $\mathscr{L}$, G and S. (The adjoint operator is defined by forming $v\mathscr{L}(u) - u\hat{\mathscr{L}}(v)$ and making this expression a divergence, i.e. of the form $\partial u/\partial x + \partial v/\partial y$ with u and v two functions[5].)

If $\mathscr{L}$ is self adjoint then $\mathscr{L} = \hat{\mathscr{L}}$. If we now choose v to be the fundamental solution $v^*(\vec{x}, \vec{\xi})$ of $\hat{\mathscr{L}}$ defined by

$$\hat{\mathscr{L}}(v^*) = \delta(\vec{x} - \vec{\xi}) \tag{4.96}$$

then equation (4.95) becomes, by equation (4.35)

$$\left.\begin{array}{c} u(\vec{\xi}) \\ cu(\vec{\xi}) \\ 0 \end{array}\right\} = \langle S(u) - p,\ \hat{G}(v^*) - q \rangle_\Gamma - \langle G(u) - q,\ \hat{S}(v^*) - p \rangle_\Gamma \quad \left\{\begin{array}{l} \vec{\xi} \in \Omega \\ \vec{\xi} \in \Gamma \\ \vec{\xi} \notin \Omega \end{array}\right. \tag{4.97}$$

notice that v^* is a function of two variables and that (86) is an identity for all values of the source or reference point $\vec{\xi}$. (4.83) is the basic equation for the direct boundary element method. If we choose $\vec{\xi}$ to be outside our problem region Ω then we have a relation between values of u on the boundary only. When these integrals are discretised in the usual boundary element way we obtain a number of equations (one for each choice of $\vec{\xi}$) involving our boundary values of u. If we choose all our $\vec{\xi}$'s to be outside Ω then *we shall have no singular integrals to evaluate.*

Another way of eliminating the integration over our problem region is to choose our weighting function v to be a homogeneous solution for the operator $\hat{\mathscr{L}}$, i.e.

$$\hat{\mathscr{L}}(v) = 0 \qquad \text{in } \Omega \tag{4.98}$$

Some care is needed here, however, as we need to choose v such that it involves another point $\vec{\xi}$ in order that we may obtain enough equations to solve our problem. In addition if our problem region extends to infinity then we need to ensure that the integrals along the part of the boundary at infinity are bounded for this solution. This condition is automatically satisfied for our choice of the fundamental solution for this function, as can be seen from its physical interpretation.[5]

Having obtained the values of u on the boundary we can re-use equation (4.87) to obtain values of the function u at any internal point $\vec{\xi}$

by now putting $\vec{\xi} \in \Omega$. We must now choose v to be the fundamental solution in order to exploit the selective property of the delta function.

Note that for these internal points we again have no singular integrals to evaluate after the boundary element discretisation.

Example 4.8

Let us now consider an example which illustrates how a non-self-adjoint problem can be solved. Consider the equation,

$$\frac{\mathrm{d}^2 u}{\mathrm{d}x^2} - \frac{\mathrm{d}u}{\mathrm{d}x} + u = 0 \tag{a}$$

with boundary conditions to be specified soon.

This equation can now be weighted by u^* to give,

$$\int_x \left(\frac{\mathrm{d}^2 u}{\mathrm{d}x^2} - \frac{\mathrm{d}u}{\mathrm{d}x} + u \right) u^* \, \mathrm{d}x = 0 \tag{b}$$

If we integrate this expression by parts twice we obtain,

$$\int \left(\frac{\mathrm{d}^2 u^*}{\mathrm{d}x^2} + \frac{\mathrm{d}u^*}{\mathrm{d}x} + u^* \right) u \, \mathrm{d}x = -[(q-u)u^*]_\Gamma + [q^* u]_\Gamma \tag{c}$$

We have to find a solution to the adjoint equation, i.e.

$$\frac{\mathrm{d}^2 u^*}{\mathrm{d}x^2} + \frac{\mathrm{d}u^*}{\mathrm{d}x} + u^* = 0 \tag{d}$$

Note that this is the same type of equation as for a damped one degree of freedom system. Hence we can take

$$u^* = \mathrm{e}^{-x/2} \left[A \sin\left(\frac{\sqrt{3}}{2} x \right) + B \cos\left(\frac{\sqrt{3}}{2} x \right) \right] \tag{e}$$

where A and B are arbitrary constants, to be determined from the boundary conditions.

Hence we can write,

$$[(q-u)u^*]_\Gamma = [q^* u]_\Gamma \tag{f}$$

Assume now that the boundary conditions for our problem are,

$$\begin{aligned} &\text{at } x = 0 \rightarrow u = 0 \\ &\text{at } x = 1 \rightarrow u = 1 \end{aligned} \tag{g}$$

Equation (f) can be written at the two points as,

$$\begin{aligned} &q_0 B \cos 1 + (q_1 - 1)e^{-1/2}[A \sin(\sqrt{3}/2) + B \cos(\sqrt{3}/2)] = q_1^* \\ &q_0 e^{-1/2}[A \sin(\sqrt{3}/2) + B \cos(\sqrt{3}/2)] + (q_1 - 1)B \cos 1 = q_1^* \end{aligned} \tag{h}$$

where

$$\begin{aligned} q^* = \frac{du^*}{dx} = -\tfrac{1}{2}e^{-x/2}\left[A \sin\left(\frac{\sqrt{3}}{2}x\right) + B \cos\left(\frac{\sqrt{3}}{2}x\right)\right] + \\ + e^{-x/2}\left[A\frac{\sqrt{3}}{2}\cos\left(\frac{\sqrt{3}}{2}x\right) - \right. \\ \left. - B\frac{\sqrt{3}}{2}\sin\left(\frac{\sqrt{3}}{2}x\right)\right] \end{aligned} \tag{i}$$

4.9 EXPLICIT FORMS FOR THE FUNDAMENTAL SOLUTION

In this section we present some explicit forms for the fundamental solution for some commonly occurring problems. These are presented in Table 4.1.

We have reverted to the simpler notation for the delta function explained in Chapter 2 and have taken our source point to be at the origin by a simple coordinate transformation.

The Mindlin solution is given in Section 4.6.

Table 4.1 SOME FUNDAMENTAL SOLUTIONS

	Equation	*Fundamental solution*
(i) *One-dimensional equations*		
Laplace	$\frac{d^2u^*}{dx^2} + \delta_0 = 0$	$u^* = r/2 \quad r = \lvert x \rvert$
Helmholtz	$\frac{d^2u^*}{dx^2} + \lambda^2 u^* + \delta_0 = 0$	$u^* = -\frac{1}{2\lambda}\sin(\lambda r)$
Wave equation	$c^2\frac{\partial^2 u^*}{\partial x^2} - \frac{\partial^2 u^*}{\partial t^2} + \delta_0\delta(t) = 0$	$u^* = \frac{1}{2c}H(ct - r)$ where H is the Heaviside or unit step function
Diffusion equation	$\frac{\partial^2 u^*}{\partial x^2} - \frac{1}{k}\frac{\partial u^*}{\partial t} + \delta_0\delta(t) = 0$	$u^* = -\frac{H(t)}{\sqrt{4\pi kt}}\exp\left(-\frac{r^2}{4kt}\right)$
Convection/ decay equation	$\frac{\partial u^*}{\partial t} + \bar{u}\frac{\partial u^*}{\partial x} + \beta u^* + \delta_0\delta(t) = 0$	$u^* = -e^{-\beta r/\bar{u}}\,\delta\left(t - \frac{r}{\bar{u}}\right)$
(ii) *Two-dimensional equations*		$r = \sqrt{x_1^2 + x_2^2}$
Laplace	$\frac{\partial^2 u^*}{\partial x_1^2} + \frac{\partial^2 u^*}{\partial x_2^2} + \delta_0 = 0$	$u^* = \frac{1}{2\pi}\ln\left(\frac{1}{r}\right)$
Helmholtz	$\frac{\partial^2 u^*}{\partial x_1^2} + \frac{\partial^2 u^*}{\partial x_2^2} + \lambda^2 u^* + \delta_0 = 0$	$u^* = \frac{1}{4i}H_0^{(2)}(\lambda r)$ where H is a Hankel function

Table 4.1 *(continued)*

	Equation	Fundamental solution
(ii) *Two-dimensional equations*		
D'Arcy	$k_1 \dfrac{\partial^2 u^*}{\partial x_1^2} + k_2 \dfrac{\partial^2 u^*}{\partial x_2^2} + \delta_0 = 0$	$u^* = \dfrac{1}{\sqrt{k_1 k_2}} \dfrac{1}{2\pi} \ln\left(\dfrac{1}{r_0}\right)$
	(orthotropic case)	where $r_0 = \left(\dfrac{x_1^2}{k_1} + \dfrac{x_2^2}{k_2}\right)^{1/2}$
Wave equation	$c^2\left(\dfrac{\partial^2 u^*}{\partial x_1^2} + \dfrac{\partial^2 u^*}{\partial x_2^2}\right) - \dfrac{\partial^2 u^*}{\partial t^2} + \delta_0 \delta(\mathrm{t}) = 0$	$u^* = -\dfrac{H(ct-r)}{2\pi c(c^2t^2 - r^2)}$
Plate equation	$\left(\dfrac{\partial^2}{\partial t^2} - \mu^2 \nabla^4\right) u^* + \delta_0 \delta(\mathrm{t}) = 0$	$u^* = +\dfrac{H(t)}{4\pi\mu} S_{\mathrm{i}}\left(\dfrac{r}{4\mu t}\right)$
	$\nabla^4 = (\nabla^2)^2$ in two dimensions	S_{i} is the integral sine function
		$S_{\mathrm{i}}(u) = -\displaystyle\int_u^\infty \dfrac{\sin v}{v}\,\mathrm{d}v$
Reduced plate equation	$k_p = \omega/\mu$	
	$(\nabla^4 - k_p^4)\, u^* + \delta_0 = 0$	$u^* = -\dfrac{1}{8\mathrm{i}k_p^2}\left(H_0^{(2)}(k_p r) - \dfrac{2\mathrm{i}}{\pi} K_0(k_p r)\right)$
		where K_0 is an elliptic function

Navier's equation (Kelvin solution)	$\frac{\partial \sigma^*_{jk}}{\partial x_j} + \delta_l = 0$ (point load in direction l)	Displacement in direction k $u^*_k = U^*_{lk} e_k = \frac{[(3-4v)\ln(1/r)\delta_{kl} + r_{,k} r_{,l}]e_l}{8\pi G(1-v)}$ Traction in direction k $p^*_k = p^*_{kl} e_l = -\frac{1}{r}\left(\frac{\mathrm{d}r}{\mathrm{d}n}[(1-2v)\delta_{lk} + 2r_{,l}r_{,k}] - (1-2v)(n_l r_{,k} - n_k r_{,l})\right)\frac{e_l}{4\pi(1-v)}$
(iii) *Three-dimensional equations*		$r = \sqrt{x_1^2 + x_2^2 + x_3^2}$
Laplace	$\frac{\partial^2 u^*}{\partial x_1^2} + \frac{\partial^2 u^*}{\partial x_2^2} + \frac{\partial^2 u^*}{\partial x_3^2} + \delta_0 = 0$	$u^* = 1/4\pi r$
Helmholtz	$\frac{\partial^2 u^*}{\partial x_1^2} + \frac{\partial^2 u^*}{\partial x_2^2} + \frac{\partial^2 u^*}{\partial x_3^2} + \lambda^2 u^* + \delta_0 = 0$	$u^* = \frac{1}{4\pi r}\exp(-\mathrm{i}\lambda r)$
D'Arcy	$k_1 \frac{\partial^2 u^*}{\partial x_1^2} + k_2 \frac{\partial^2 u^*}{\partial x_2^2} + k_3 \frac{\partial^2 u^*}{\partial x_3^2} + \delta_0 = 0$	$u^* = \frac{1}{\sqrt{k_1 k_2 k_3}} \frac{1}{4\pi} \frac{1}{r_0}$ where $r_0 = \sqrt{\frac{x_1^2}{k_1} + \frac{x_2^2}{k_2} + \frac{x_3^2}{k_3}}$
Wave equation	$c^2 \nabla^2 u^* - \frac{\partial^2 u^*}{\partial t^2} + \delta_0 \delta(\mathrm{t}) = 0$	$u^* = \frac{\delta(t-(r/c))}{4\pi r}$

Table 4.1 *(continued)*

	Equation	Fundamental solution
(iii) *Three-dimensional equations*		
Navier's equation (isotropic, homogeneous, Kelvin solution)	$\frac{\partial \sigma_{jk}^*}{\partial x_j} + \delta_l = 0$	Displacement in direction k $u_k^* = U_{lk}^* e_l$ $u_{lk}^* = \frac{1}{16\pi G(1-\nu)}\left(\frac{3-4\nu}{r}\tilde{\delta}_{lk} + \frac{\partial r}{\partial x_l}\frac{\partial r}{\partial x_k}\right)$
	(point load in direction l)	Traction in direction k $p_i^* = p_{ji}^* e_j$ $p_j^* = p_{ij}^* e_i$ $= -\frac{1}{8\pi(1-\nu^2)r^2}\left(\frac{\partial r}{\partial n}[(1-2\nu)\tilde{\delta}_{ij} + 3r_{,i}r_{,j}] + (1-2\nu)(njr_{,i} - n_i r_{,j})\right)e_j$

REFERENCES

1. Titchmarsh, E. C., *Eigenfunction Expansions Associated with Second-Order Differential Equations* (2 vols) Clarendon Press (1958)
2. Synge, J. L., *The Hypercircle in Mathematical Physics* Cambridge University Press (1957)
3. Mindlin, R. D., 'Force at a Point in the Interior of a Semi-Infinite Solid', *Physics*, **7**, p. 145 May (1936)
4. Garabedian, P. R., *Partial Differential Equations* John Wiley and Sons (1964)
5. Brebbia, C. A. and Walker, S., 'Simplified Boundary Elements for Radiation Problems', *Res. Note, Appl. Math. Modelling*, **2**, No. 2, June (1978)

5 Elastostatics

5.1 INTRODUCTION

In this chapter we will study elastostatics problems assuming that we have linear material behaviour and neglecting any changes in orientation of the body due to displacements. In what follows we will use the indicial notation for brevity.

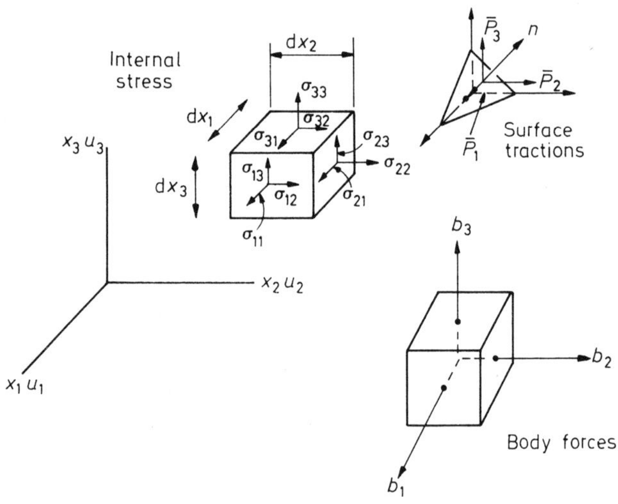

Figure 5.1 Internal stress, body forces and surface tractions

The governing equations can be written in terms of the stress field components (see Figure 5.1) σ_{ij} as,

$$\begin{aligned}
&\frac{\partial\sigma_{11}}{\partial x_1}+\frac{\partial\sigma_{12}}{\partial x_2}+\frac{\partial\sigma_{13}}{\partial x_3}+b_1=0\\
&\frac{\partial\sigma_{21}}{\partial x_1}+\frac{\partial\sigma_{22}}{\partial x_2}+\frac{\partial\sigma_{23}}{\partial x_3}+b_2=0\\
&\frac{\partial\sigma_{31}}{\partial x_1}+\frac{\partial\sigma_{32}}{\partial x_2}+\frac{\partial\sigma_{33}}{\partial x_3}+b_3=0
\end{aligned}\tag{5.1}$$

where $\sigma_{21}=\sigma_{12},\sigma_{31}=\sigma_{13},\sigma_{32}=\sigma_{23}$.

Equations (5.1) can be written in indicial notation, simply as

$$\frac{\partial\sigma_{ij}}{\partial x_j}+b_i=0 \qquad i=1,2,3 \quad j=1,2,3 \tag{5.2}$$

The prescribed surface force intensities or tractions on the surface Γ_2 of the body are denoted by $\bar{p}_1,\bar{p}_2,\bar{p}_3$ (Figure 5.1). Equilibrium on the boundary requires the satisfaction of the following boundary conditions:

$$\begin{aligned}
p_1&=\sigma_{11}n_1+\sigma_{12}n_2+\sigma_{13}n_3=\bar{p}_1\\
p_2&=\sigma_{21}n_1+\sigma_{22}n_2+\sigma_{23}n_3=\bar{p}_2\\
p_3&=\sigma_{31}n_1+\sigma_{32}n_2+\sigma_{33}n_3=\bar{p}_3
\end{aligned}\tag{5.3}$$

where n_1,n_2,n_3 are the direction cosines of the normal with respect to the x_1,x_2,x_3 axes, i.e.

$$n_i=\cos(n,x_i) \tag{5.4}$$

Equations (5.3) can be written in indicial form as,

$$p_i=\sigma_{ij}n_j=\bar{p}_i \qquad i=1,2,3 \quad j=1,2,3 \tag{5.5}$$

The displacement field is defined by the three displacement components u_1,u_2,u_3. Let Γ_1 denote the portion of the boundary on which displacements are prescribed. The displacement constraints there are,

$$u=\bar{u}_1,\quad u_2=\bar{u}_2,\quad u_3=\bar{u}_3 \tag{5.6}$$

or

$$u_i=\bar{u}_i.$$

The $\bar{u}_i$ are the prescribed values. Note that the total surface Γ of the body is equal to $\Gamma_1+\Gamma_2$.

The strains at any point are defined by the ε_{ij} components of the strain tensor, which in indicial form are,

$$\varepsilon_{ij} = \frac{1}{2}\left(\frac{\partial u_i}{\partial x_j} + \frac{\partial u_j}{\partial x_i}\right) \qquad i = 1, 2, 3 \quad j = 1, 2, 3 \tag{5.7}$$

The strains and stresses are related by the constitutive relations which for an isotropic body are,

$$\sigma_{ij} = \lambda \tilde{\delta}_{ij} \frac{\partial u_k}{\partial x_k} + \mu\left(\frac{\partial u_i}{\partial x_j} + \frac{\partial u_j}{\partial x_i}\right) \tag{5.8}$$

where $\tilde{\delta}_{ij}$ is the Kronecker delta and λ, μ are Lamé's constants which can also be expressed in terms of the modulus of elasticity E and Poisson's ratio ν by,

$$\lambda = \frac{E}{(1+\nu)(1-2\nu)}, \qquad \mu = \frac{E}{2(1+\nu)} \tag{5.9}$$

Sometimes μ is written as G, which is the shear modulus.

In certain problems different effects, such as temperature, can be included as initial stresses or strains. If the effects are considered as initial stresses we have,

$$\begin{aligned} \sigma_{ij} &= \lambda \tilde{\delta}_{ij} \frac{\partial u_k}{\partial x_k} + \mu\left(\frac{\partial u_i}{\partial x_j} + \frac{\partial u_j}{\partial x_i}\right) + \sigma_{ij}^0 \\ &= \sigma_{ij}^t + \sigma_{ij}^0 \end{aligned} \tag{5.10}$$

where σ_{ij} are now elastic stresses, equal to total stresses σ_{ij}^t and the σ_{ij}^0 components. Note that σ_{ij}^0 for a thermally isotropic material is

$$\sigma_{ij}^0 = -\lambda\, \varepsilon_{ij}^0 \tilde{\delta}_{ij} \tag{5.11}$$

For a temperature problem the ε_{ij}^0 are

$$\varepsilon_{ij}^0 = \alpha T \tag{5.12}$$

where α is the dilatation coefficient and T the difference in temperature. The effect of initial stresses or strains can be analysed in the same way as body forces.

5.2 WEIGHTED RESIDUAL STATEMENTS

Following the same ideas as developed in the previous chapters we can write an extended weighted residual equation taking into consideration the equilibrium equations (5.2) and the two types of boundary conditions (5.5) and (5.6). This statement is,

$$\int_{\Omega}\left(\frac{\partial\sigma_{jk}}{\partial x_j}+b_k\right)u_k^*\,\mathrm{d}\Omega=\int_{\Gamma_2}(p_k-\bar{p}_k)u_k^*\,\mathrm{d}\Gamma+\int_{\Gamma_1}(\bar{u}_k-u_k)p_k^*\,\mathrm{d}\Gamma \tag{5.13}$$

where u_k^*, p_k^* are the displacements and surface forces corresponding to the weighting field:

$$p_k^*=n_j\sigma_{jk}^* \tag{5.14}$$

The strain–displacement relationship (5.7) and the constitutive equations (5.8) are assumed to apply for both, the approximating and the weighting fields.

We can now integrate (5.13) by parts which gives,

$$-\int_{\Omega}\sigma_{jk}\varepsilon_{jk}^*\,\mathrm{d}\Omega+\int_{\Omega}b_ku_k^*\,\mathrm{d}\Omega=-\int_{\Gamma_2}\bar{p}_ku_k^*\,\mathrm{d}\Gamma-\int_{\Gamma_1}p_ku_k^*\,\mathrm{d}\Gamma+$$
$$+\int_{\Gamma_1}(\bar{u}_k-u_k)p_k^*\,\mathrm{d}\Gamma \tag{5.15}$$

Let us consider for generality that,

$$\sigma_{jk}=\sigma_{jk}^t+\sigma_{jk}^0 \tag{5.16}$$

where σ^t are the total stresses and σ^0 the initial stresses.

Taking this into consideration we can write (5.15) as

$$-\int_{\Omega}\sigma_{jk}^t\varepsilon_{jk}^*\,\mathrm{d}\Omega-\int_{\Omega}\sigma_{jk}^0\varepsilon_{jk}^*\,\mathrm{d}\Omega+\int_{\Omega}b_ku_k^*\,\mathrm{d}\Omega$$
$$=-\int_{\Gamma_2}\bar{p}_ku_k^*\,\mathrm{d}\Gamma-\int_{\Gamma_1}p_ku_k^*\,\mathrm{d}\Gamma+\int_{\Gamma_1}(\bar{u}_k-u_k)p_k^*\,\mathrm{d}\Gamma \tag{5.17}$$

Integrating by parts again the first term in (5.17) and taking into consideration the constitutive relations (5.10) we have,

$$\int_{\Omega} \frac{\partial \sigma_{jk}^*}{\partial x_j} u_k \, d\Omega - \int_{\Omega} \sigma_{jk}^0 \varepsilon_{jk}^* \, d\Omega + \int_{\Omega} b_k u_k^* \, d\Omega$$

$$= -\int_{\Gamma_2} \bar{p}_k u_k^* \, d\Gamma - \int_{\Gamma_1} p_k u_k^* \, d\Gamma + \int_{\Gamma_1} \bar{u}_k p_k^* \, d\Gamma + \int_{\Gamma_2} u_k p_k^* \, d\Gamma \tag{5.18}$$

For temperature effects for instance this equation becomes (see equations (5.11) and (5.12)),

$$\int_{\Omega} \frac{\partial \sigma_{jk}^*}{\partial x_j} u_k \, d\Omega + \int_{\Omega} \lambda \alpha T \tilde{\delta}_{jk} \varepsilon_{jk}^* \, d\Omega + \int_{\Omega} b_k u_k^* \, d\Omega +$$

$$+ \int_{\Gamma_2} \bar{p}_k u_k^* \, d\Gamma + \int_{\Gamma_1} p_k u_k^* \, d\Gamma = \int_{\Gamma_2} u_k p_k^* \, d\Gamma + \int_{\Gamma_1} \bar{u}_k p_k^* \, d\Gamma \tag{5.19}$$

This shows that the initial stress field can be treated in a similar way to the body force field b_k. In addition to being used for temperature and other problems the initial stress fields are important because they can be applied to introduce the effect of non-linearities into the formulation. These two effects do not produce any internal unknowns and they relate to the boundary values in a similar way to the p term in the Poisson equation.

The problem is to find a solution such that

$$\partial \sigma_{jk}^* / \partial x_j = 0 \tag{5.20}$$

In this way the first integral in equation (5.19) disappears which reduces the problem to a boundary problem. We need to find the solution to this homogeneous equation in order to apply (5.19) without having to integrate the first term in equation (5.19) over the Ω domain, which would produce internal unknowns.

5.3 FUNDAMENTAL SOLUTION

A way of applying (5.19) is to use the fundamental solution for the elasticity problem, i.e. the solution corresponding to the equation

$$\frac{\partial \sigma_{jk}^*}{\partial x_j} + \delta_l^i = 0 \tag{5.21}$$

where δ_l^i is the Dirac delta function and represents a unit load at i acting in *one* of the x_l directions. This type of solution will produce for each direction l the following equation (the initial stress term has not been considered for simplicity):

$$u_l^i + \int_{\Gamma_1} \bar{u}_k p_k^* \, d\Gamma + \int_{\Gamma_2} u_k p_k^* \, d\Gamma$$
$$= \int_{\Omega} b_k u_k^* \, d\Omega + \int_{\Gamma_1} p_k u_k^* \, d\Gamma + \int_{\Gamma_2} \bar{p}_k u_k^* \, d\Gamma \qquad (5.22)$$

u_l^i represents the displacement at i in the l direction (see Figure 5.2), u_k^* and p_k^* are the displacements and tractions respectively due to a unit concentrated load at the point i acting in the l direction. If we consider unit forces acting at i in the three directions, equation (5.22) can then be written

$$u_l^i + \int_{\Gamma_1} \bar{u}_k p_{lk}^* \, d\Gamma + \int_{\Gamma_2} u_k p_{lk}^* \, d\Gamma$$
$$= \int_{\Omega} b_k u_{lk}^* \, d\Omega + \int_{\Gamma_1} p_k u_{lk}^* \, d\Gamma + \int_{\Gamma_2} \bar{p}_k u_{lk}^* \, d\Gamma \qquad (5.23)$$

where p_{lk}^* and u_{lk}^* represent the tractions and displacements in the k direction due to unit forces acting in the direction l.

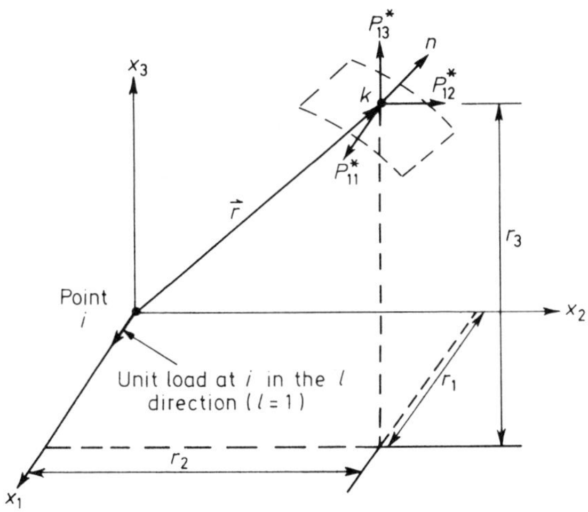

Figure 5.2 Notation. k is a point on the surface; p_{lk} are surface forces at k due to a unit load at i acting in the l direction ($l = 1$)

The fundamental solution used for a three-dimensional isotropic body is the Kelvin solution and corresponds to a concentrated force acting at a point in the infinite elastic space. For the displacements it can be written,

$$u_{lk}^* = \frac{1}{16\pi G(1-\nu)}\left(\frac{(3-4\nu)\tilde{\delta}_{lk}}{r} + \frac{\partial r}{\partial x_l}\frac{\partial r}{\partial x_k}\right) \tag{5.24}$$

where we interpret the derivatives as

$$\frac{\partial r}{\partial x_k} = \frac{r_k}{r} \tag{5.25}$$

where r_k are the projections defined in Figure 5.2.

The traction components corresponding to the Kelvin solution are

$$p_{lk}^* = -\frac{1}{8\pi(1-\nu^2)r^2}\left[\frac{\partial r}{\partial n}\left((1-2\nu)\tilde{\delta}_{lk} + 3\frac{\partial r}{\partial x_l}\frac{\partial r}{\partial x_k}\right) + \right.$$

$$\left. + (1-2\nu)\left(\frac{\partial r}{\partial x_l}n_k - \frac{\partial r}{\partial x_k}n_l\right)\right] \tag{5.26}$$

n is the normal to the surface of the body.

Equation (5.23) can be specialised for a boundary point in the same way as the potential problem. Due to the singularities existing in the left-hand side integrals, we obtain a coefficient such that,

$$c_i u_l^i + \int_{\Gamma_1} \bar{u}_k p_{lk}^* \, d\Gamma + \int_{\Gamma_2} u_k p_{lk}^* \, d\Gamma$$

$$= \int_{\Omega} b_k u_{lk}^* \, d\Omega + \int_{\Gamma_1} p_k u_{lk}^* \, d\Gamma + \int_{\Gamma_2} \bar{p}_k u_{lk}^* \, d\Gamma \tag{5.27}$$

The c_i coefficient is equal to $\frac{1}{2}$ for a smooth boundary but generally will be different from this value. Fortunately explicit calculation of this value is not necessary as it can be obtained using the rigid body motions in a way similar to that explained in Chapter 2 for potential problems. We will come back to this in Section 5.4.

5.4 SOURCE APPROACH

Boundary solutions can also be expressed using a source of dipole type formulation. Here we will discuss only the source one, the other formulation can equally well be deduced as was done for potential problems.

Consider for a moment that all the boundary is of the type Γ_1 with a value of the displacements equal to $\hat{u}_k$ (displacements for an external field). This field also has some components of surface tractions that can be calculated from $\hat{u}_k$ and are called $\hat{p}_k$. Hence equation (5.23) becomes,

$$u_l^i + \int_\Gamma u_k p_{lk}^* \, \mathrm{d}\Gamma = \int_\Omega b_k u_{lk}^* \, \mathrm{d}\Omega + \int_\Gamma p_k u_{lk}^* \, \mathrm{d}\Gamma \tag{5.28}$$

From the special form of the divergence theorem for this case we can write,

$$\int_\Gamma \hat{u}_k p_{lk}^* \, \mathrm{d}\Gamma = \int_\Gamma \hat{p}_k u_{lk}^* \, \mathrm{d}\Gamma \tag{5.29}$$

Thus equation (5.28) can be written,

$$u_l^i = \int_\Gamma (p_k - \hat{p}_k) u_{lk}^* \, \mathrm{d}\Gamma + \int_\Gamma b_k u_{lk}^* \, \mathrm{d}\Gamma \tag{5.30}$$

where $p_k - \hat{p}_k$ is defined as the ϕ_k source and formula (5.30) can be written

$$u_l^i = \int_\Gamma \phi_k u_{lk}^* \, \mathrm{d}\Gamma + \int_\Gamma b_k u_{lk}^* \, \mathrm{d}\Gamma \tag{5.31}$$

The value of the force components can now be deduced as

$$p_l^i = \int_\Gamma \phi_k p_{lk}^* \, \mathrm{d}\Gamma + \int_\Gamma b_k p_{lk}^* \, \mathrm{d}\Gamma \tag{5.32}$$

Equation (5.31) presents a singularity when the point is taken to the boundary. Its final version gives for a smooth boundary,

$$p_l^i = -\frac{1}{2}\phi_l^i + \int_\Gamma \phi_k u_{lk}^* \, \mathrm{d}\Gamma + \int_\Gamma b_k u_{lk}^* \, \mathrm{d}\Gamma \tag{5.33}$$

Formulae (5.32) and (5.33) can be used to solve the problem expressed in terms of σ sources. If we start with a Γ_2 type of boundary one will arrive at the dipole formulation. Although we have started here assuming that the boundary was Γ_1 type, we can solve in terms of known p's and u's using the above formulae.

5.5 MATRIX FORMULATION

We can now work with matrices instead of using the indicial notation. In order to do so we can define **u** as the displacement, **p** as the traction and **b** as the body force vectors, such that

$$\mathbf{u} = \begin{Bmatrix} u_1 \\ u_2 \\ u_3 \end{Bmatrix}, \quad \mathbf{p} = \begin{Bmatrix} p_1 \\ p_2 \\ p_3 \end{Bmatrix}, \quad \mathbf{b} = \begin{Bmatrix} b_1 \\ b_2 \\ b_3 \end{Bmatrix} \tag{5.34}$$

At a point i these displacements are called $\mathbf{u}^i$. In addition we can define the following two matrices:

$$\mathbf{u}^* = \begin{bmatrix} u_{11}^* & u_{12}^* & u_{13}^* \\ u_{21}^* & u_{22}^* & u_{23}^* \\ u_{31}^* & u_{32}^* & u_{33}^* \end{bmatrix}, \quad \mathbf{p}^* = \begin{bmatrix} p_{11}^* & p_{12}^* & p_{13}^* \\ p_{21}^* & p_{22}^* & p_{23}^* \\ p_{31}^* & p_{32}^* & p_{33}^* \end{bmatrix} \tag{5.35}$$

where the u_{lk} and p_{lk} are the displacements and forces in the k directions due to a unit force at the point i under consideration, acting in the l direction.

Let us first write formula (5.27) before any boundary conditions are applied, i.e.

$$c_i u_l^i + \int_\Gamma p_{lk}^* u_k \, d\Gamma = \int_\Gamma u_{lk}^* p_k \, d\Gamma + \int_\Omega u_{lk}^* b_k \, d\Omega \tag{5.36}$$

This equation can now be expressed in matrix form as follows,

$$c_i \mathbf{u}_i + \int_\Gamma \mathbf{p}^* \mathbf{u} \, d\Gamma = \int_\Gamma \mathbf{u}^* \mathbf{p} \, d\Gamma + \int_\Omega \mathbf{u}^* \mathbf{b} \, d\Omega \tag{5.37}$$

This formulation is valid for a point i. Note that $\mathbf{p}^*$, $\mathbf{u}^*$ and $\mathbf{b}$ are known and c_i will soon be determined. The unknowns are the values of **u** and **p** over the boundary.

BOUNDARY ELEMENTS

We will now assume that the boundary is divided into elements. These boundary elements can be constant, linear, quadratic or higher order and they can be triangles or quadrilaterals. Figure 5.3(a) shows the body divided into a series of constant elements. For these elements the unknowns are assumed to be at the centre of the element and to have a constant value over the element. The elements of Figure 5.3(b) are such that the functions vary linearly over them. Finally in Figure 5.3(c) the variation of displacements and stresses over each element is quadratic. In addition the geometry of the element is described by a quadratic function.

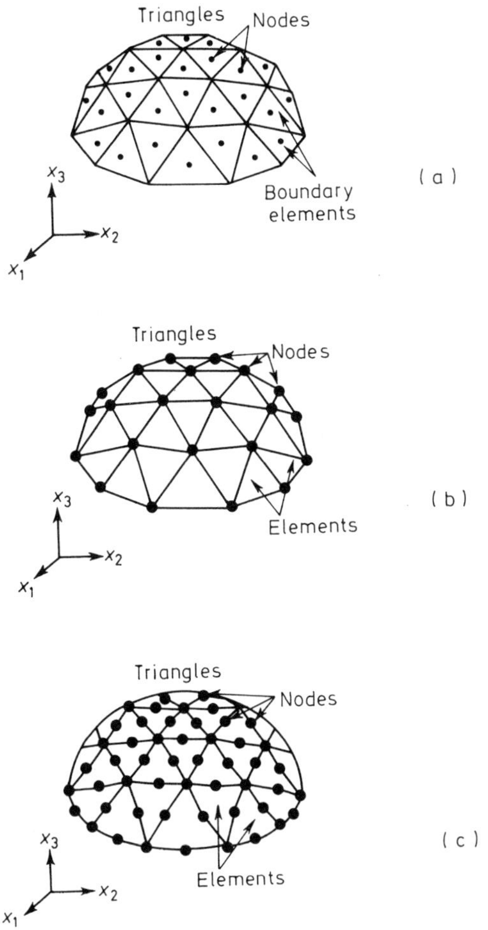

Figure 5.3 Three-dimensional bodies divided into elements: (a) *constant;* (b) *linear; and* (c) *quadratic*

Over each element the functions for u and p can be approximated using the following interpolation functions:

$$\mathbf{u} = \mathbf{\Phi}^{\mathrm{T}}\mathbf{u}^{\mathrm{n}} = \begin{bmatrix} \boldsymbol{\phi}^{\mathrm{T}} & \cdot & \cdot \\ \cdot & \boldsymbol{\phi}^{\mathrm{T}} & \cdot \\ \cdot & \cdot & \boldsymbol{\phi}^{\mathrm{T}} \end{bmatrix} \mathbf{u}^{\mathrm{n}} \tag{5.38}$$

$$\mathbf{p} = \boldsymbol{\psi}^{\mathrm{T}}\mathbf{p}^{\mathrm{n}} = \begin{bmatrix} \boldsymbol{\psi}^{\mathrm{T}} & \cdot & \cdot \\ \cdot & \boldsymbol{\psi}^{\mathrm{T}} & \cdot \\ \cdot & \cdot & \boldsymbol{\psi}^{\mathrm{T}} \end{bmatrix} \mathbf{p}^{\mathrm{n}} \tag{5.39}$$

Note that the functions for u and p are generally different. For consistence it may be better to take the functions for **p** of one order less than those for **u**. The interpolation functions are the ones discussed in Chapter 3. $\mathbf{u}^{\mathrm{n}}$, $\mathbf{p}^{\mathrm{n}}$ are the nodal displacements and tractions.

We can substitute the above functions into (5.37) and obtain the following equation for the point i:

$$c_i\mathbf{u}_i + \sum_{j=1}^{N}\left(\int_{\Gamma_j} \mathbf{p}^*\mathbf{\Phi}^{\mathrm{T}}\,\mathrm{d}\Gamma\right)\mathbf{u}^{\mathrm{n}} = \sum_{j=1}^{N}\left(\int_{\Gamma_j} \mathbf{u}^*\boldsymbol{\psi}^{\mathrm{T}}\,\mathrm{d}\Gamma\right)\mathbf{p}^{\mathrm{n}} + \\ + \sum_{j=1}^{M}\int_{\Omega_{\mathrm{m}}} \mathbf{u}^*\mathbf{b}\,\mathrm{d}\Omega\Bigg) \tag{5.40}$$

where the summation from $j = 1$ to N indicates summation over the N elements on the surface and Γ_j is the surface of the j element. The summation from $j = 1$ to M is carried out over the internal cells and Ω_{m} is the volume of each of them.

The integrals in equation (5.40) are usually solved numerically and the functions $\mathbf{\Psi}$ and $\mathbf{\Phi}$ are expressed in some of the homogeneous system of coordinates described in Chapter 3. The integrals need then to be written in terms of the homogeneous system, to be called ξ, η. Hence

$$\mathrm{d}\Gamma = (\text{absolute value of } |\mathbf{G}|)\;\mathrm{d}\xi\,\mathrm{d}\eta \tag{5.41}$$

where **G** has been defined previously.

Due to the difficulties in integrating equation (5.40) it is usual to employ a numerical integration scheme. Hence formula (5.40) becomes,

$$c_i\mathbf{u}_i + \sum_{j=1}^{N}\left(\sum_{r=1}^{L}|\mathbf{G}|\,w_r\,(\mathbf{p}^*\boldsymbol{\phi}^{\mathrm{T}}(\xi,\eta))_r\right)\mathbf{u}^{\mathrm{n}}$$

$$= \sum_{j=1}^{N}\left(\sum_{r=1}^{L}|\mathbf{G}|\,w_r(\mathbf{u}^*\,\boldsymbol{\psi}^{\mathrm{T}}(\xi,\eta))\right)\mathbf{p}^{\mathrm{n}} + \sum_{j=1}^{M}\left(\sum_{r=1}^{T}|\mathbf{J}|\,w_r(\mathbf{u}^*\mathbf{b})_r\right) \tag{5.42}$$

J is the Jacobian for the three-dimensional transformation, and w_r are the numerical integration weighting coefficients.

Note that (5.42) gives a set of equations for node i which can be written as,

$$c_i\mathbf{u}_i + [\hat{\mathbf{h}}_{i1}\,\hat{\mathbf{h}}_{i2}\,\ldots\,\hat{\mathbf{h}}_{ii}\,\ldots\,\hat{\mathbf{h}}_{ir}]\begin{Bmatrix}\mathbf{u}_1\\ \mathbf{u}_2\\ \vdots\\ \mathbf{u}_i\\ \vdots\\ \mathbf{u}_r\end{Bmatrix} \tag{5.43}$$

$$= [\mathbf{g}_{i1}\,\mathbf{g}_{i2}\,\ldots\,\mathbf{g}_{ii}\,\ldots\,\mathbf{g}_{ir}]\begin{Bmatrix}\mathbf{p}_1\\ \mathbf{p}_2\\ \vdots\\ \mathbf{p}_i\\ \vdots\\ \mathbf{p}_r\end{Bmatrix} + \mathbf{b}_i$$

where $\mathbf{u}_j$ and $\mathbf{p}_j$ are the unknowns at nodes j. $\hat{\mathbf{h}}_{ij}$ and $\mathbf{g}_{ij}$ are the interaction coefficients relating node i. with all the nodes on the surface of the body. Note that with the exception of constants elements, more than one element will contribute to the $\hat{\mathbf{h}}_{ij}$ and $\mathbf{g}_{ij}$ submatrices. (They are 3×3 for the three-dimensional case and 2×2 for two dimensions.)

SYSTEM EQUATIONS

We can write a matrix equation such as (5.43) for each of the nodes under consideration. Writing them together we have,

$$\begin{bmatrix} \mathbf{h}_{11} & \mathbf{h}_{22} & \cdots & \mathbf{h}_{1i} & \cdots & \mathbf{h}_{1r} \\ \mathbf{h}_{21} & \mathbf{h}_{22} & \cdots & \mathbf{h}_{2i} & \cdots & \mathbf{h}_{2r} \\ \vdots & \vdots & & \vdots & & \vdots \\ \mathbf{h}_{i1} & \mathbf{h}_{i2} & \cdots & \mathbf{h}_{ii} & \cdots & \mathbf{h}_{ir} \\ \vdots & \vdots & & \vdots & & \vdots \\ \mathbf{h}_{r1} & \mathbf{h}_{r2} & \cdots & \mathbf{h}_{ri} & \cdots & \mathbf{h}_{rr} \end{bmatrix} \begin{Bmatrix} \mathbf{u}_1 \\ \mathbf{u}_2 \\ \vdots \\ \mathbf{u}_i \\ \vdots \\ \mathbf{u}_r \end{Bmatrix}$$

$$= \begin{bmatrix} \mathbf{g}_{11} & \mathbf{g}_{12} & \cdots & \mathbf{g}_{1i} & \cdots & \mathbf{g}_{1r} \\ \mathbf{g}_{21} & \mathbf{g}_{22} & \cdots & \mathbf{g}_{2i} & \cdots & \mathbf{g}_{2r} \\ \vdots & \vdots & & \vdots & & \vdots \\ \mathbf{g}_{i1} & \mathbf{g}_{i2} & \cdots & \mathbf{g}_{ii} & \cdots & \mathbf{g}_{ir} \\ \vdots & \vdots & & \vdots & & \vdots \\ \mathbf{g}_{r1} & \mathbf{g}_{r2} & \cdots & \mathbf{g}_{ri} & \cdots & \mathbf{g}_{rr} \end{bmatrix} \begin{Bmatrix} \mathbf{p}_1 \\ \mathbf{p}_2 \\ \vdots \\ \mathbf{p}_i \\ \vdots \\ \mathbf{p}_r \end{Bmatrix} + \begin{Bmatrix} \mathbf{b}_1 \\ \mathbf{b}_2 \\ \vdots \\ \mathbf{b}_i \\ \vdots \\ \mathbf{b}_r \end{Bmatrix} \tag{5.44}$$

where the submatrices $\mathbf{h}_{ii}$ on the diagonal are,

$$\mathbf{h}_{ii} = \hat{\mathbf{h}}_{ii} + \mathbf{c}_i \tag{5.45}$$

Equations (5.44) can be written as,

$$\mathbf{H}\,\mathbf{U} = \mathbf{G}\,\mathbf{P} + \mathbf{B} \tag{5.46}$$

We now have to apply the boundary conditions in this system which are of two types: (i) $\mathbf{u} = \overline{\mathbf{u}}_i$ on Γ_1; and (ii) $\mathbf{p}_i = \overline{\mathbf{p}}_i$ on Γ_2. If the displacements are known we find the tractions. This implies that the system of equations (5.46) can be reordered in such a way that all the unknowns are written on the left-hand side in an $\mathbf{X}$ vector. The final result can be written as

$$\mathbf{A}\mathbf{X} = \mathbf{F} \tag{5.47}$$

The body force vector is included in $\mathbf{F}$.

Note that we have not yet determined the c coefficients of equation (5.45). They can be obtained implicitly by consideration of rigid body movements. Let us assume we have a unit rigid body displacement in any one direction; equation (5.46) then becomes,

$$\mathbf{H}\mathbf{I}_l = \mathbf{O} \tag{5.48}$$

where $\mathbf{I}_l$ is a vector defining a unit rigid displacement in the direction 'l'. Hence the diagonal terms of $\mathbf{H}$ are simply

$$\mathbf{h}_{ii} = \sum_{i \neq j} \mathbf{h}_{ij} \tag{5.49}$$

which means that the c coefficients do not need to be determined explicitly.

Once equations (5.47) have been solved we know the stresses and displacements over the whole surface of the body. We can now compute the stresses and displacements at any internal point.

STRESSES AND DISPLACEMENTS INSIDE THE BODY

The displacements at any point inside the body are given by equation (5.28), i.e.

$$\mathbf{u} = \int_\Gamma \mathbf{u}^* \, \mathbf{p} \, \mathrm{d}\Gamma - \int_\Gamma \mathbf{p}^* \, \mathbf{u} \, \mathrm{d}\Gamma + \int_\Omega \mathbf{u}^* \, \mathbf{b} \, \mathrm{d}\Omega \tag{5.50}$$

or for each l component at that point,

$$u_l = \int_\Gamma u_{lk}^* \, p_k \, \mathrm{d}\Gamma - \int_\Gamma p_{lk}^* \, u_k \, \mathrm{d}\Gamma + \int_\Omega u_{lk}^* \, b_k \, \mathrm{d}\Omega \tag{5.51}$$

The internal stresses for an isotropic medium can be obtained by applying equation (5.8)

$$\sigma_{ij} = \lambda \, \tilde{\delta}_{ij} \frac{\partial u_l}{\partial x_l} + \mu \left(\frac{\partial u_i}{\partial x_j} + \frac{\partial u_j}{\partial x_i} \right) \tag{5.52}$$

Substituting (5.51) into (5.52) we obtain,

$$\begin{aligned}\sigma_{ij} = {} & \int_\Gamma \left[\lambda \, \tilde{\delta}_{ij} \frac{\partial u_{lk}^*}{\partial x_l} + \mu \left(\frac{\partial u_{ik}^*}{\partial x_j} + \frac{\partial u_{jk}^*}{\partial x_i} \right) \right] p_k \, \mathrm{d}\Gamma - \\ & - \int_\Gamma \left[\lambda \, \tilde{\delta}_{ij} \frac{\partial p_{lk}^*}{\partial x_l} + \mu \left(\frac{\partial p_{ik}^*}{\partial x_j} + \frac{\partial p_{jk}^*}{\partial x_i} \right) \right] u_k \, \mathrm{d}\Gamma + \\ & + \int_\Omega \left[\lambda \, \tilde{\delta}_{ij} \frac{\partial u_{lk}^*}{\partial x_l} + \mu \left(\frac{\partial u_{ik}^*}{\partial x_j} + \frac{\partial u_{jk}^*}{\partial x_i} \right) \right] b_k \, \mathrm{d}\Omega \end{aligned} \tag{5.53}$$

We can write this as

$$\sigma_{ij} = \int_\Gamma D_{kij} p_k \, d\Gamma - \int_\Gamma S_{kij} u_k \, d\Gamma + \int_\Omega D_{kij} b_k \, d\Omega \qquad (5.54)$$

where for three dimensions,

$$D_{kij} = \frac{1}{r^2}\left[(1-2\nu)\left(\tilde{\delta}_{ki}\frac{\partial r}{\partial x_j} + \tilde{\delta}_{kj}\frac{\partial r}{\partial x_i} - \tilde{\delta}_{ij}\frac{\partial r}{\partial x_k}\right) + \right.$$
$$\left. + 3\frac{\partial r}{\partial x_i}\frac{\partial r}{\partial x_j}\frac{\partial r}{\partial x_k}\right]\frac{1}{8\pi(1-\nu)} \qquad (5.55)$$

$$S_{kij} = \frac{2\mu}{3r}\left\{3\frac{\partial r}{\partial n}\left[(1-2\nu)\tilde{\delta}_{ij}\frac{\partial r}{\partial x_k} + \nu\left(\tilde{\delta}_{ik}\frac{\partial r}{\partial x_j} + \tilde{\delta}_{jk}\frac{\partial r}{\partial x_i}\right) - \right.\right.$$
$$\left. - 5\frac{\partial r}{\partial x_i}\frac{\partial r}{\partial x_j}\frac{\partial r}{\partial x_k}\right] + 3\nu\left(n_i\frac{\partial r}{\partial x_j}\frac{\partial r}{\partial x_k} + n_j\frac{\partial r}{\partial x_i}\frac{\partial r}{\partial x_k}\right) +$$
$$+ (1-2\nu)\left(3n_k\frac{\partial r}{\partial x_i}\frac{\partial r}{\partial x_j} + n_j\tilde{\delta}_{ik} + n_i\tilde{\delta}_{jk}\right) -$$
$$\left. - (1-4\nu)n_k\tilde{\delta}_{ij}\right\}\frac{1}{8\pi(1-\nu)} \qquad (5.56)$$

The derivatives of r with respect to the coordinates are shown in Figure 5.4 for a two-dimensional case for simplicity.

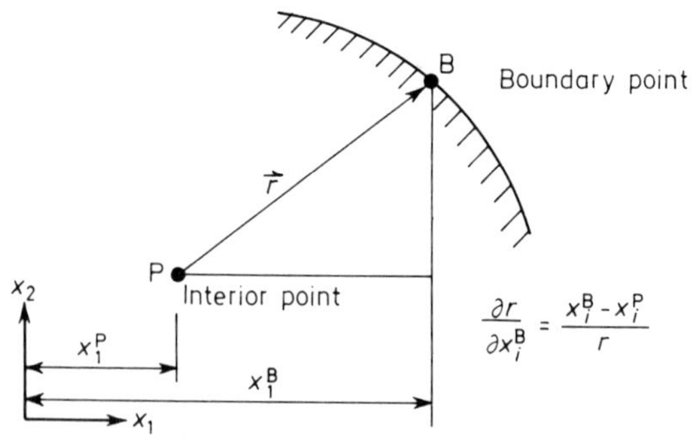

Figure 5.4 Derivatives of r: P, interior point; B, boundary point

Example 5.1[1]

As an example of the application of the above techniques, consider the problem of determining the stress concentration for a spherical cavity. Cruse used an equation similar to (5.27) as the starting point of his analysis. Then discretising the surface of the problem region (Figure 5.5) he was able to represent the region using quadratic curved surface elements. (The triangular elements shown in Figure 5.5 are just degenerate elements with essentially the same form.) To represent the surface curvature Cruse used quadratic shape functions to map the elements and as he used quadratic interpolation functions on each element his formulation was of the isoparametric form. (see Chapter 3.)

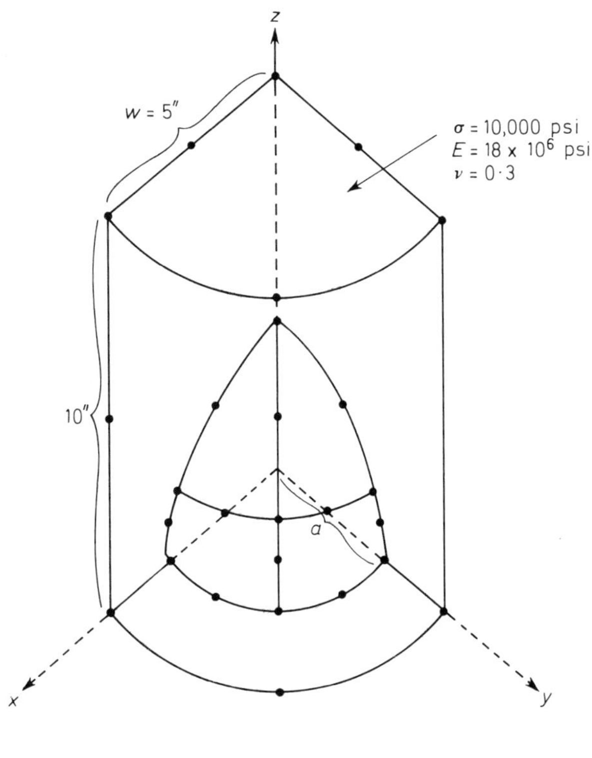

Figure 5.5 Boundary element discretisation for spherical cavity 6 elements (cylinder 2, cavity 4), 28 nodes

By exploiting the symmetry of the problem he was able to represent the surface using only six elements.

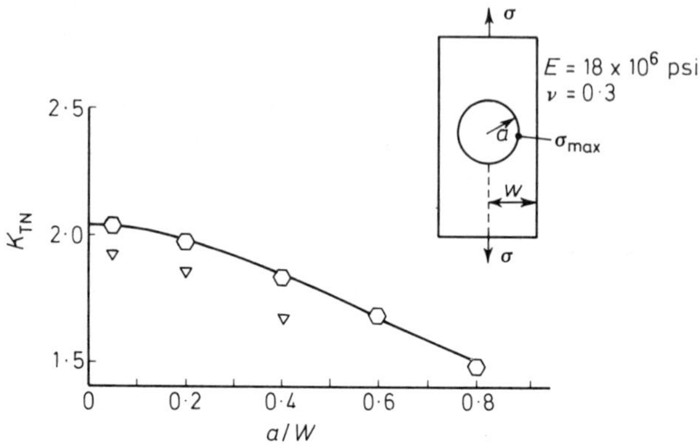

Figure 5.6 Stress concentration for a spherical cavity. $K_{TN} = \sigma_{max}/\bar{\sigma}$ *where* $\bar{\sigma} = \sigma/[1 - (a/w)^2]$*; the full curve is from Peterson and the data points are from boundary integral equations:* ⬡*, quadratic;* ∇*, linear*

The results for the stress concentration shown in Figure 5.6 show good agreement with published results and also compare the results using the quadratic isoparametric boundary elements, with the corresponding results obtained using the simpler linear boundary elements.

Example 5.2[1]

In the same paper referred to above, Cruse has studied the problem of crack propagation by looking at the stresses around a circular crack, again using quadratic boundary elements. Figure 5.7 shows the boundary element discretisation used to model a quarter of a semi-cylinder with an interior or surface crack. Only one quarter was considered because of the symmetry of the problem.

The numerical results for the circular crack are given in Figure 5.8; for the buried crack, linear and quadratic variations are compared. The quadratic shape function representation more accurately reflects the displacements in the body. The quarter points referred to in the figure are mid-side nodes one quarter of the element side length away

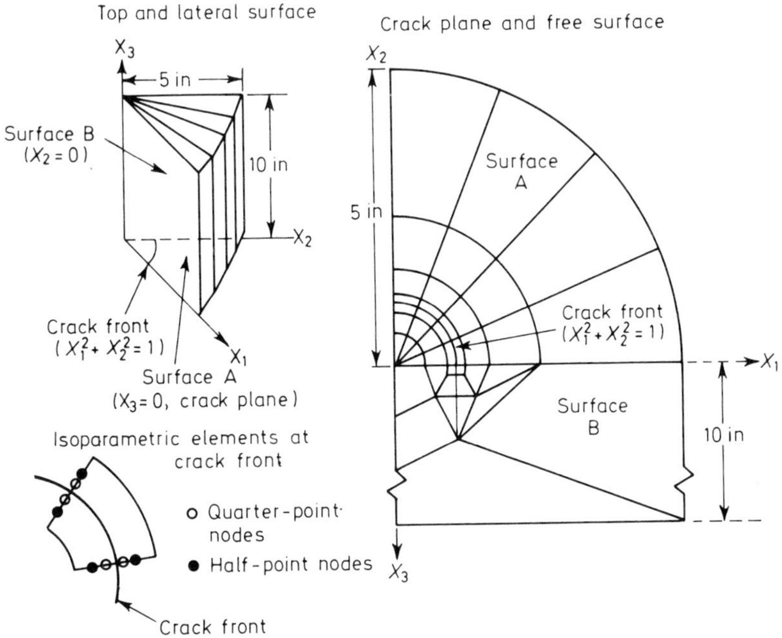

Figure 5.7 Boundary element discretisation for circular crack

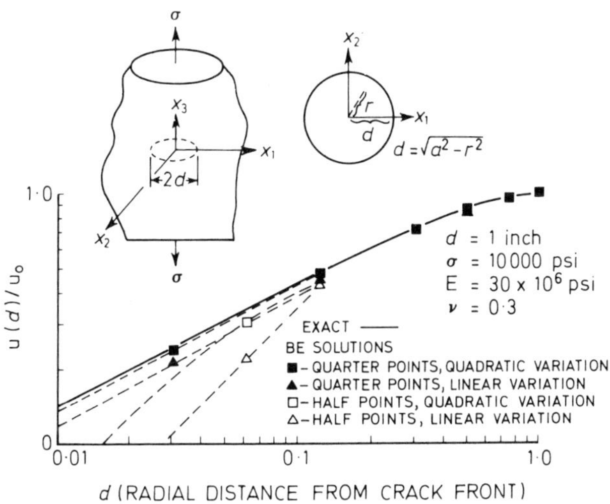

Figure 5.8 Displacement variation for a buried circular crack

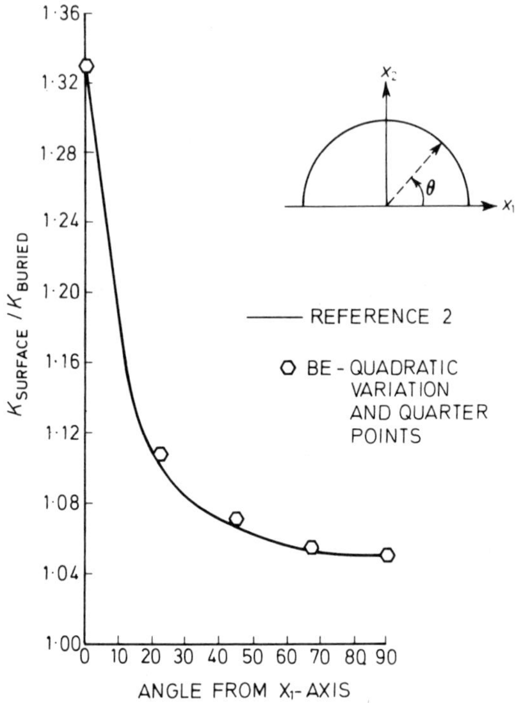

Figure 5.9 Stress intensity factor variation for a semicircular crack

from the crack for the two rows of elements adjacent to the crack tip.

Figure 5.9 shows the variation of the surface crack stress intensity factor using crack opening displacements as described in Cruse and Meyers[2]. The results are compared with those obtained in Cruse and Meyers[2] using linear elements; it is clear that linear variation models are sufficiently accurate to determine the stress intensity.

Example 5.3[3]

At the intersection of two thick cylinders stress concentration problems may occur (Figure 5.10). In Figures 5.10 and 5.11 one particular type of joint is illustrated. Due to symmetry only one eighth of the joint needs to be studied.

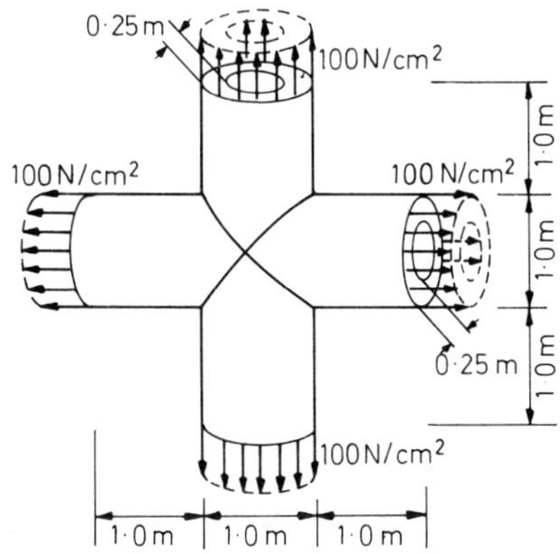

Figure 5.10 Joint for two thick cylinders

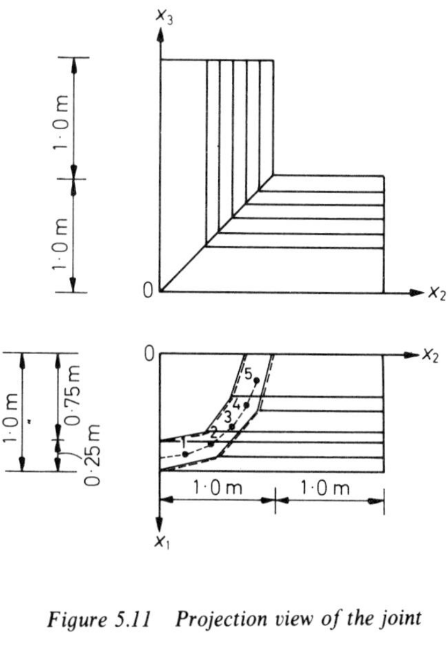

Figure 5.11 Projection view of the joint

The system was discretised using 76 constant triangular boundary elements as in Figure 5.12 and stresses were calculated at five points at the intersection (Figure 5.11). Results for normal stresses in the z direction are shown in Figure 5.13. They indicate the distribution of stresses at these points.

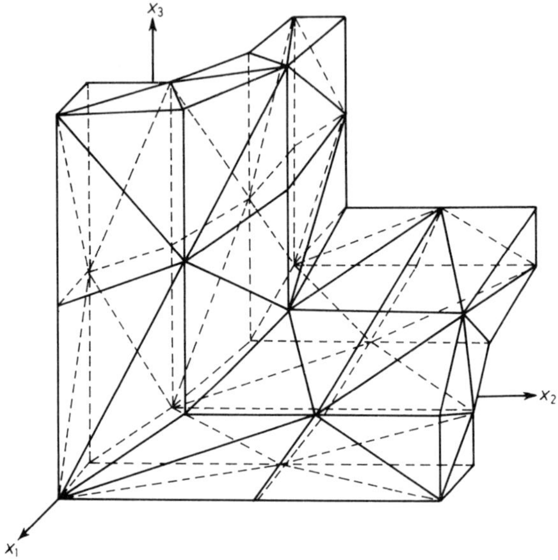

Figure 5.12 Discretisation of the joint

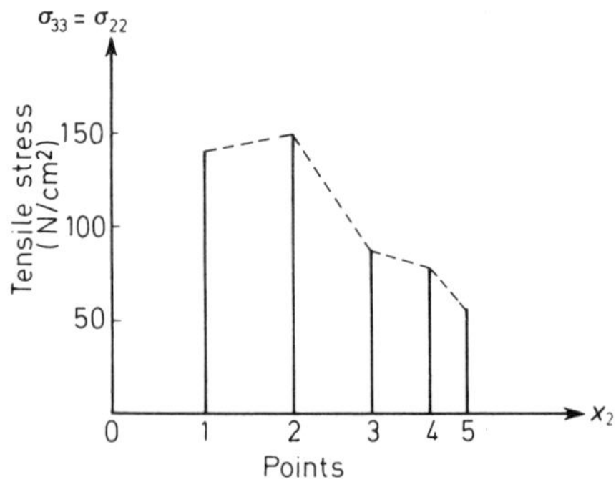

Figure 5.13 Results for the joint at each point

5.6 TWO-DIMENSIONAL ELASTICITY

Let us now apply the boundary element method in two-dimensional elasticity. The fundamental solution of the governing equations, i.e.

$$\sigma^*_{jk,j} = \delta_i \quad i = 1, 2; j = 1, 2 \tag{5.57}$$

is as following (plane strain)

$$u^*_{lk} = \frac{1}{8\pi\mu(1-\nu)}\left[(3-4\nu)\ln\left(\frac{1}{r}\right)\tilde{\delta}_{lk} + \frac{\partial r}{\partial x_l}\frac{\partial r}{\partial x_r}\right]$$

$$p^*_{lk} = \frac{1}{4\pi(1-\nu)r}\left[\frac{\partial r}{\partial n}\left((1-2\nu)\tilde{\delta}_{lk} + 2\frac{\partial r}{\partial x_k}\frac{\partial r}{\partial x_l}\right) + \right. \tag{5.58}$$

$$\left. + (1-2\nu)\left(\frac{\partial r}{\partial x_l}n_k - \frac{\partial r}{\partial x_k}n_l\right)\right]$$

where p_{lk} and u_{lk} represent the tractions and displacements in the k direction due to a unit force in the l direction.

The elements of p^* and u^* can be written in matrix form as follows,

$$\mathbf{p}^* = \begin{bmatrix} p^*_{11} & p^*_{12} \\ p^*_{21} & p^*_{22} \end{bmatrix}, \quad \mathbf{u}^* = \begin{bmatrix} u^*_{11} & u^*_{12} \\ u^*_{21} & u^*_{22} \end{bmatrix} \tag{5.59}$$

The displacements, tractions and body force vectors are now

$$\mathbf{u} = \begin{Bmatrix} u_1 \\ u_2 \end{Bmatrix}, \quad \mathbf{p} = \begin{Bmatrix} p_1 \\ p_2 \end{Bmatrix}, \quad \mathbf{b} = \begin{Bmatrix} b_1 \\ b_2 \end{Bmatrix} \tag{5.60}$$

The starting equation can be written in matrix form as for three-dimensional elasticity (equation (5.37)), i.e.

$$c_i \mathbf{u}_i + \int_\Gamma \mathbf{p}^* \mathbf{u} \, d\Gamma = \int_\Gamma \mathbf{u}^* \mathbf{p} \, d\Gamma + \int_\Omega \mathbf{u}^* \mathbf{b} \, d\Omega \tag{5.61}$$

The rest of the matrix operations are similar to those described in Section 5.5 for the three-dimensional case, but here the boundary integrals are line integrals and the body force terms are obtained by integrating over area instead of volume of internal elements.

Once the unknowns are known over all the surface one can find the internal displacements and stresses at any internal point using,

$$\mathbf{u} = \int_\Gamma \mathbf{u}^* \mathbf{p} \, \mathrm{d}\Gamma - \int_\Gamma \mathbf{p}^* \mathbf{u} \, \mathrm{d}\Gamma + \int_\Omega \mathbf{u}^* \mathbf{b} \, \mathrm{d}\Omega \tag{5.62}$$

$$\sigma_{ij} = \int_\Gamma D_{kij} p_k \, \mathrm{d}\Gamma - \int_\Gamma S_{kij} u_k \, \mathrm{d}\Gamma + \int_\Omega D_{kij} u_k \, \mathrm{d}\Omega \tag{5.63}$$

where,

$$D_{kij} = \frac{1}{r}\left[(1-2\nu)\left(\tilde{\delta}_{ki}\frac{\partial r}{\partial x_j} + \tilde{\delta}_{kj}\frac{\partial r}{\partial x_i} - \tilde{\delta}_{ij}\frac{\partial r}{\partial x_k}\right) + \right.$$
$$\left. + 2\frac{\partial r}{\partial x_i}\frac{\partial r}{\partial x_j}\frac{\partial r}{\partial x_k}\right]\frac{1}{4\pi(1-\nu)} \tag{5.64}$$

$$S_{kij} = -\frac{\mu}{r}\left\{2\frac{\partial r}{\partial n}\left[(1-2\nu)\tilde{\delta}_{ij}\frac{\partial r}{\partial x_k} + \nu\left(\tilde{\delta}_{ik}\frac{\partial r}{\partial x_j} + \tilde{\delta}_{jk}\frac{\partial r}{\partial x_i}\right) - \right.\right.$$
$$\left. -4\frac{\partial r}{\partial x_i}\frac{\partial r}{\partial x_j}\frac{\partial r}{\partial x_k}\right] + 2\nu\left(n_i\frac{\partial r}{\partial x_j}\frac{\partial r}{\partial x_k} + n_j\frac{\partial r}{\partial x_i}\frac{\partial r}{\partial x_k}\right) +$$
$$+ (1-2\nu)\left(2n_k\frac{\partial r}{\partial x_i}\frac{\partial r}{\partial x_j} + n_j\tilde{\delta}_{ik} + n_i\tilde{\delta}_{jk}\right) -$$
$$\left. -(1-4\nu)n_k\tilde{\delta}_{ij}\right\}\frac{1}{4\pi(1-\nu)} \tag{5.65}$$

As previously the integrals can be calculated using numerical integration formulae, or analytically for very simple cases.

Note that in order to pass from plane strain to two-dimensional plane stress analysis one has to substitute the values of Poisson's ratio and modulus of elasticity by,

$$E \Leftrightarrow E\left(1 - \frac{\nu^2}{(1+\nu)^2}\right); \; \nu \Leftrightarrow \nu/(1+\nu) \tag{5.66}$$

The thermal strain coefficient is now

$$\alpha' \Leftrightarrow \alpha\left(\frac{1+\nu}{1+2\nu}\right)$$

Example 5.4

The problem of crack propagation in two dimensions has been examined by Lachat.[4] This situation is depicted in Figure 5.14 and the

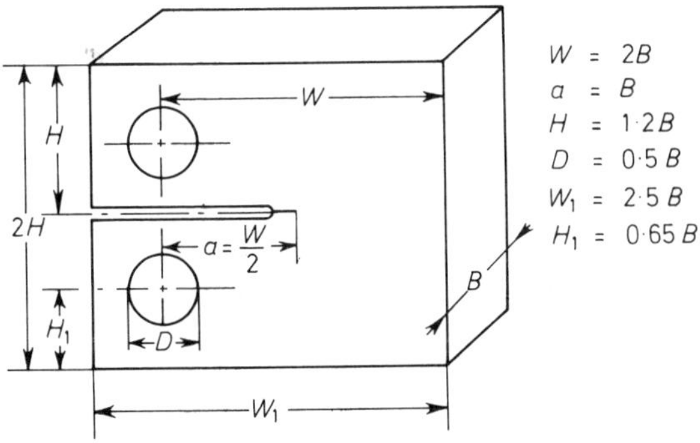

Figure 5.14 Crack geometry

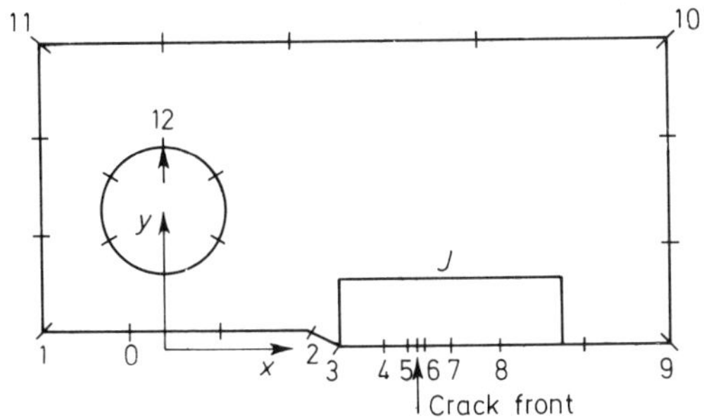

Figure 5.15 Boundary element discretisation

discretisations used are shown in Figures 5.15 and 5.16. Cubic boundary elements were used throughout and a traction was applied at node 12 in the direction indicated. The propagation of the crack resulting from the applied traction was found and compared against finite element calculations; the results compare very well.

The stress intensity factor was found for different specimens by evaluating Rice's integral along the contour Γ (see Figure 5.15) and experimental results were compared against finite and boundary element solutions. The agreement between the three solutions was satisfactory.

Cruse[1, 5] tackled a similar problem using a modified fundamental solution which contained the full elastic representation of a flat,

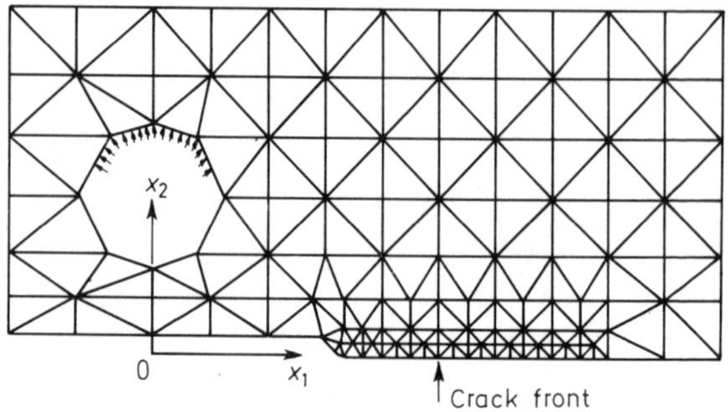

Figure 5.16 Finite element discretisation

traction free crack. This fundamental solution was obtained using the method of images and complex mapping described in Chapter 4. In this way the formulation automatically included the presence of the crack without having to model the crack surface geometry. This method however, can only be applied in two-dimensional problems.

Example 5.5[6]

The stress concentration around holes in an anisotropic sheet has been examined by Krenk[6] using the boundary element technique. It is usual to model such holes by a distribution of point sources corresponding to the fundamental solution. The fundamental solution is usually obtained in complex form in terms of the complex roots of the characteristic equation;[7] however, Krenk points out that this makes their interpretation very difficult. Krenk therefore uses a real form of this solution and avoids the problems of double roots associated with the isotropic case.

An example of the method was chosen to be the case of two circular holes in an infinite anisotropic sheet loaded at infinity (Figure 5.17). Dislocations with an unknown density are distributed on the two contours to make the surface tractions zero along the inside. Rigid body rotations are eliminated by imposing an additional constraint on the integral equation deduced earlier in the paper.

Numerical calculations were carried out for five different materials: isotropic and orthotropic materials with properties giving rise to double roots of the characteristic equation and others giving only

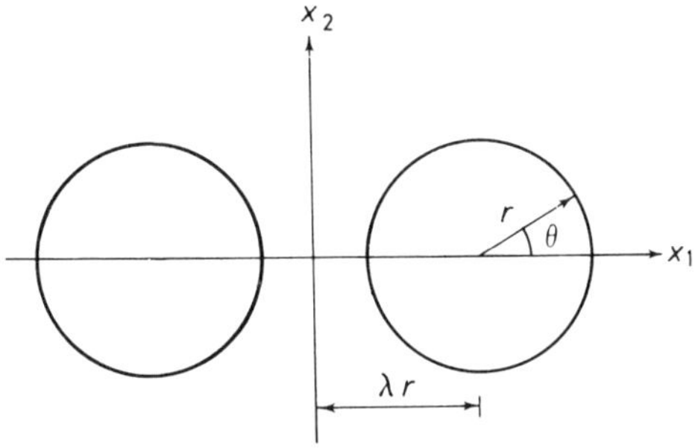

Figure 5.17 Two circular holes in an anisotropic sheet

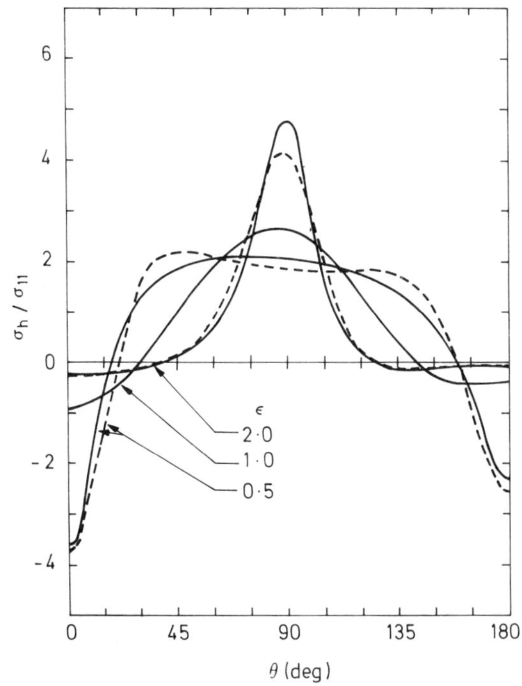

Figure 5.18 Hoop stress (σ_n) around circular hole for tension in x_1 direction

simple roots. Different orientations of the axes of orthotropy were also tried.

Figures 5.18 and 5.19 represent the hoop stresses for longitudinal

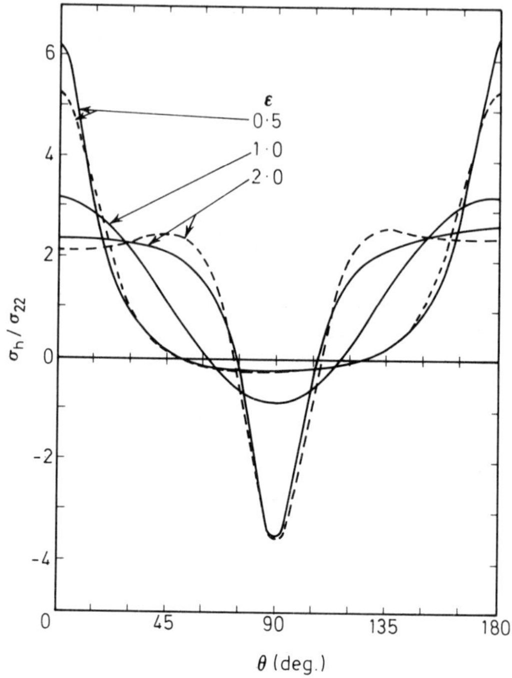

Figure 5.19 Hoop stress (σ_h) around circular hole for tension in x_2 direction

and transverse tension respectively. λ was taken to be equal to 1·5. In both diagrams the dashed lines represent the case where we have double roots with

$$\kappa = \frac{1}{2}\sqrt{E_1 E_2}\left(\frac{1}{G} - \frac{2\nu_{12}}{E_1}\right) = 1 \tag{a}$$

where

$$E_1 = 1/a_{1111} \tag{b}$$

$$E_2 = 1/a_{2222} \tag{c}$$

$$G = 1/4a_{1212} \tag{d}$$

$$\nu_{12}/E_1 = -a_{1122} \tag{e}$$

where the a's are the usual components of the tensor relating stress to strain in an anisotropic solid. We have written

$$\varepsilon = \sqrt[4]{E_1/E_2} \tag{f}$$

From the figures we can see that the interaction between the circles is small.

Example 5.6[8]

Lachat[8] examined the case of a rubber sheet with embedded steel wires under plane longitudinal strain (Figure 5.20). He used cubic

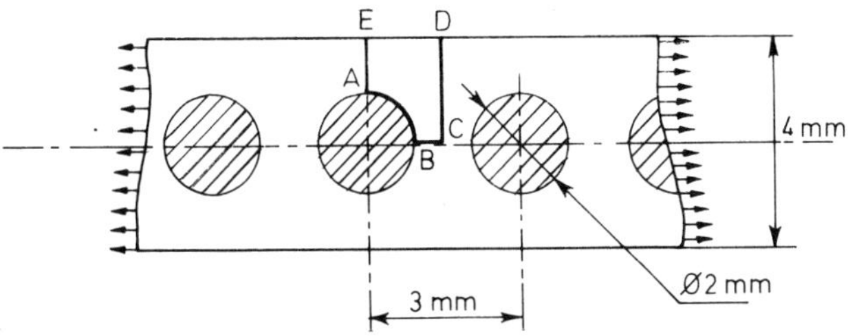

Figure 5.20 Sheet of rubber with embedded steel wires

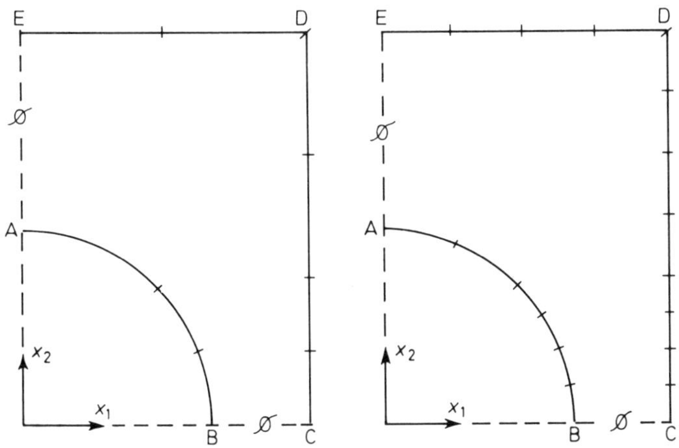

Figure 5.21 Discretisation for boundary element analysis: (a) *coarse discretisation (9 segments);* (b) *fine discretisation (18 segments)*

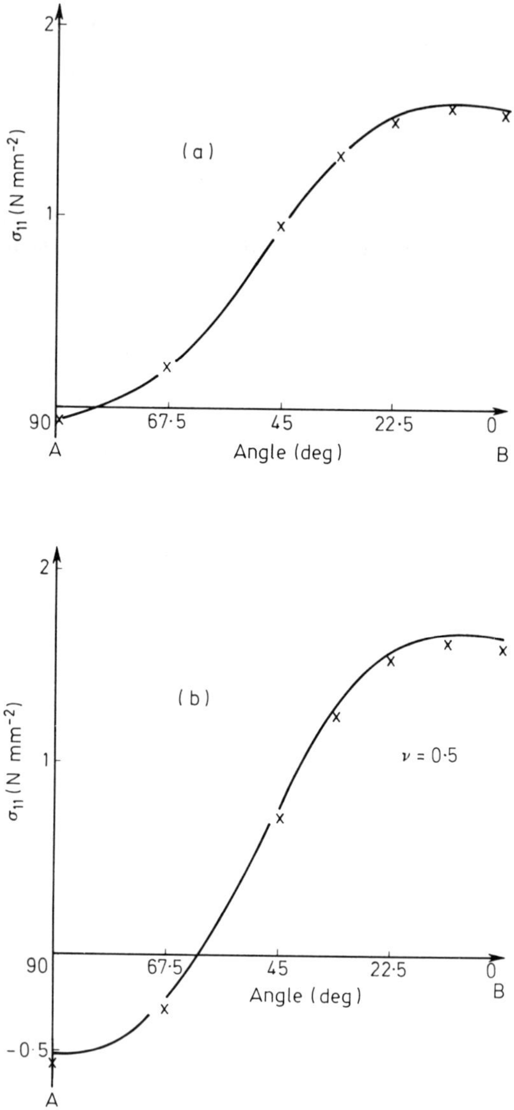

Figure 5.22 *Variation of* σ_{11} *along the curve AB:* (a) $\nu = 0.45$; (b) $\nu = 0.5$

elements and used two different boundary element discretisations: one with 9 segments and the other with 18 segments for comparison (Figure 5.21). By the symmetry of the problem only the area ABCDE of Figure 5.20 was considered.

For the purposes of the analysis the wires were taken to be perfectly rigid. The overall imposed strain was $1{\cdot}667 \times 10^{-2}$ in the x_1 direction. The modulus of elasticity was taken to be 20 N mm^{-2} and the two values of Poisson's ratio $\nu = 0{\cdot}45$ and $\nu = 0{\cdot}5$ were considered.

For $\nu = 0{\cdot}45$ the stresses σ_{11} calculated using the two meshes were virtually indistinguishable (Figure 5.22(a)). The stars represent the results for the coarse mesh and the solid line shows the values obtained using the finer, 18 element mesh.

For $\nu = 0{\cdot}5$ (Figure 5.22(b)) this difference was greater, up to 5 of the maximum stress. However, there is very little difference between the displacements.

Note. The finite element displacement method is incapable of treating the case $\nu = 0{\cdot}5$.

REFERENCES

1. Cruse, T. A. and Wilson, R. B., 'Advanced Applications of Boundary Integral Equation Methods', *Nuclear Engng Design*, **46**, 223–34 (1978)
2. Cruse, T. A. and Meyers, G. J., 'Two dimensional fracture mechanics analysis', *J. Struct. Div, Proc. ASCE*, **103**, 309–20 (1977)
3. Brebbia, C. A. and Nakaguma, R., 'Applications of Boundary Elements in the Analysis of Offshore Structures', *Proc. Brazil Offshore/77* Rio de Janeiro, Pentech Press (1978)
4. Boissenot, J. M., Lachat, J. C. and Watson, J., "Étude par Équations Intégrals D'une Éprouvette C. T. 15', *Rev. Physique Appl.*, **9**, 611, July (1974)
5. Cruse, T. A., 'Two Dimensional BIE Fracture Mechanics Analysis', *Proc. 1st Int. Seminar on Recent Advances in Boundary Element Methods, Southampton University, July 1978*, Pentech Press (1978)
6. Krenk, S., 'Stress Concentration Around Holes in Anisotropic Sheets', *Proc. 1st Int. Seminar on Recent Advances in Boundary Element, Methods, Southampton University, July 1978*, Pentech Press (1978)
7. Lekhnitskii, S. G., *Theory of Elasticity of an Anisotropic Elastic Body*, Holden-Day Inc., San Francisco (1963)
8. Lachat, J. C., 'A Further Development of the Boundary Integral Technique for Elastostatics', *Ph.D. Thesis*, University of Southampton, Dept of Civil Engineering, February (1975)
9. Peterson, R. E. *Stress Concentration Factors*, Wiley, New York (1953)

BIBLIOGRAPHY

Cruse, T. A., 'Application of the Boundary Integral Equation Solution Method in Solid Mechanics' in *Variational Methods in Engineering*, Eds C. Brebbia and H. Tottenham, 2 vols, Southampton University Press (1973)

Rizzo, F. J. 'An Integral Approach to Boundary Value Problems of Classical Elastostatics', *Q. Appl. Math.*, **25**, No. 83 (1967)

Rizzo, F. J. and Shippy, R., 'A Formulation and Solution Procedure for the General Non-Homogeneous Elastic Inclusion Problem', *Int. J. Solids Struct.*, **4**, 1161 (1968)

Alarcon, E., Brebbia C., and Dominguez, J., 'The Boundary Element Method in Elasticity', *Int. J. Mech. Sci.* **20**, 625–39 (1978)

Muskheliskvili, N. I., *Some Basic Problems of the Mathematical Theory of Elasticity*, Noordhoff, Holland (1953)

6 Time-Dependent and Non-Linear Problems

6.1 INTRODUCTION

Up to now we have been considering time-independent problems; many situations of physical interest involve the solution of problems with some time component, be it a vibration problem where a steady state solution is sought, the solution to a problem where the response to some transient effect is required or the case where the properties of the material itself are time dependent.

Previous analyses of time-dependent problems using the boundary element method have been solved by either integrating numerically in time in some finite difference fashion or by eliminating the time variation by the use of a Laplace transform[1, 2, 3, 4], notably Cruse.[1] The analysis of a simple vibrational problem has been attempted by Hutchinson[5] effectively, though not explicitly, using the Fourier decomposition method.

In this chapter the relatively unexplored topic of the application of boundary elements to problems with both material and certain types of geometric non-linearities will also be discussed. The final section will include a discussion of the recent applications of boundary elements to plasticity problems.[6]

6.2 TRANSFORM METHODS

In many time-dependent problems involving governing equations which are linear in the time variable and have time-independent coefficients it is possible to remove the time dependence in the equations by taking a Laplace (L) or Fourier transform ($\mathscr{F}$). The resulting equations may then be solved in the transform space with the usual boundary element method. Equations involving the time variable are usually hyperbolic (or parabolic) in type and as such are unsuitable for solution by the boundary element technique, the transformed equations, however, are in many cases elliptic.

Once the equations are solved in the transformed space the original variable involving the time dependence may be recovered by inverting the transformation numerically. The type of transform used, either Laplace (L) or Fourier ($\mathscr{F}$) will depend on the type of solution sought. We shall illustrate this point by considering a simple 'forced vibration' problem.

Consider the second-order differential equation,

$$m\frac{d^2u}{dt^2}+c\frac{du}{dt}+ku = Fe^{i\omega t} \tag{6.1}$$

This equation represents the forced vibration problem shown in Figure 6.1. Here m is the mass of the system, c the damping, k the stiffness and $Fe^{i\omega t}$ is some forcing function with angular frequency ω; u is the displacement of the mass.

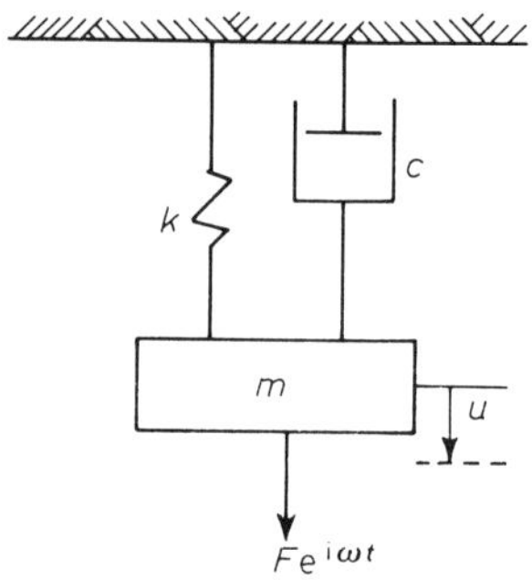

Figure 6.1 Forced vibration of a damped system

Equation (6.1) has a general solution consisting of the following two parts.

(i) The *complementary* function obtained by considering the homogeneous form of the equation,

$$m\frac{d^2u}{dt^2}+c\frac{du}{dt}+ku = 0 \tag{6.2}$$

This solution is given by,

$$u = A\,\mathrm{e}^{\alpha_+ t} + B\,\mathrm{e}^{\alpha_- t} \tag{6.3}$$

where

$$\alpha_{\pm} = -\frac{c}{2m} \pm \frac{\sqrt{c^2 - 4m}}{2m} \tag{6.4}$$

A and B are constants determined by the initial displacement u_0 and velocity $\dot{u}_0$. By considering u and its time derivative at time $t = 0$, we find that

$$A = \frac{\dot{u}_0 - \alpha_- u_0}{\alpha_+ - \alpha_-} \quad \text{and} \quad B = \frac{\dot{u}_0 - \alpha_+ u_0}{\alpha_- - \alpha_+}$$

This is the 'transient' part of the solution. In general we shall use the Laplace transform to obtain solutions of this kind.

(ii) The *particular* integral or solution obtained by postulating a solution of the type,

$$u = U\,\mathrm{e}^{\mathrm{i}\omega t} \tag{6.5}$$

This solution corresponds to a motion with the same frequency as the forcing function and for a stable system will be the 'steady state' response to the forcing function $F\,\mathrm{e}^{\mathrm{i}\omega t}$. This eventual steady state will be independent of the initial values u_0 and $\dot{u}_0$.

For our equation this solution is given by

$$U = \frac{1}{-\omega^2 m + \mathrm{i}\omega c + k} F \tag{6.6}$$

We see that U depends on the exciting frequency ω and the properties of the system m, c and k. In general we shall use the Fourier transform to obtain solutions of this kind.

The general solution of (6.1) is hence

$$u = \frac{\dot{u}_0 - \alpha_- u_0}{\alpha_+ - \alpha_-}\,\mathrm{e}^{\alpha_+ t} + \frac{\dot{u}_0 - \alpha_+ u_0}{\alpha_- - \alpha_+}\,\mathrm{e}^{\alpha_- t} + \frac{F^{\mathrm{i}\omega t}}{-\omega^2 m + \mathrm{i}\omega c + k} \tag{6.7}$$

where α_+ and α_- are given by (6.4).

6.3 FOURIER TRANSFORMS

If $f(t)$ is any piecewise differential function with a finite number of finite discontinuities (at a discontinuity $f(t)$ is taken to have the mean value of the limits obtained by approaching the discontinuity from either side), then we may define the Fourier transform $F(\omega)$ of $f(t)$ by

$$F(\omega) = \mathscr{F}[f(t)] \equiv \frac{1}{\sqrt{2\pi}} \int_{-\infty}^{\infty} f(t)\, \mathrm{e}^{-\mathrm{i}\omega t}\, \mathrm{d}t \tag{6.8}$$

Then $f(t)$ is given by

$$f(t) = \mathscr{F}^{-1}[F(\omega)] \equiv \frac{1}{\sqrt{2\pi}} \int_{-\infty}^{\infty} F(\omega)\, \mathrm{e}^{\mathrm{i}\omega t}\, \mathrm{d}\omega \tag{6.9}$$

if the integrals in (6.8) and (6.9) exist.[7]

Differentiating (6.9) with respect to t we obtain,

$$\frac{\partial f}{\partial t} = \frac{1}{\sqrt{2\pi}} \int_{-\infty}^{\infty} (\mathrm{i}\omega F(\omega))\, \mathrm{e}^{\mathrm{i}\omega t}\, \mathrm{d}\omega \tag{6.10}$$

i.e.

$$\mathscr{F}[\partial f/\partial t] = \mathrm{i}\omega\, F(\omega) \tag{6.11}$$

Similarly we can show that,

$$\mathscr{F}[\partial^2 f/\partial t^2] = -\omega^2 F(\omega) \tag{6.12}$$

Hence taking the Fourier transform of an equation linear in time (with time-independent coefficients) will replace the differentials by powers of $\mathrm{i}\omega$.

It only remains for us to define the Fourier convolution (represented by $*$) of two functions $f(t)$ and $g(t)$ by,

$$\begin{aligned} f*g &= \frac{1}{2\pi} \int_{-\infty}^{\infty} f(t-\tau)\, g(\tau)\, \mathrm{d}\tau \\ &= \frac{1}{2\pi} \int_{-\infty}^{\infty} f(\tau)\, g(t-\tau)\, \mathrm{d}\tau \\ &= \frac{1}{2\pi} \int_{-\infty}^{\infty} f(t+\tau)\, g(\tau)\, \mathrm{d}\tau \\ &= \frac{1}{2\pi} \int_{-\infty}^{\infty} f(\tau)\, g(t-\tau)\, \mathrm{d}\tau \end{aligned} \tag{6.13}$$

One important property of this convolution is that,

$$\mathscr{F}[f * g] = F(\omega)\, G(\omega) \tag{6.14}$$

Similarly

$$\mathscr{F}^{-1}[F * G] = f(t)\, g(t) \tag{6.15}$$

Example 6.1

Consider the equation,

$$m\frac{\mathrm{d}^2 u}{\mathrm{d}t^2} + c\frac{\mathrm{d}u}{\mathrm{d}t} + ku = f(t) \tag{a}$$

taking the Fourier transform throughout this equation we obtain,

$$-\omega^2 m\, U(\omega) + \mathrm{i}\omega c\, U(\omega) + k\, U(\omega) = F(\omega) \tag{b}$$

where

$$U(\omega) = \mathscr{F}[u(t)] \tag{c}$$

and

$$F(\omega) = \mathscr{F}[f(t)] \tag{d}$$

Hence,

$$U(\omega) = \frac{F(\omega)}{-\omega^2 m + \mathrm{i}\omega c + k} \tag{e}$$

The particular integral corresponding to the steady state solution of (a) is given by,

$$u(t) = \mathscr{F}^{-1}[U(\omega)] = \frac{1}{\sqrt{2\pi}} \int_{-\infty}^{\infty} \frac{F(\omega)\, \mathrm{e}^{\mathrm{i}\omega t}}{-\omega^2 m + \mathrm{i}\omega c + k}\, \mathrm{d}\omega \tag{f}$$

The integral is usually evaluated by contour integration, although in the application to boundary element methods a numerical scheme must be used in most cases.[8]

NUMERICAL INVERSION OF THE FOURIER TRANSFORM

If the forcing function $f(t)$ is sampled at N points at time intervals Δt up to a time T then we shall obtain a sequence of values $\{f_k\}$, $k = 0, 1,$

2, . . . , $N-1$ and (6.8) may be represented approximately by

$$F_k = \frac{1}{T}\sum_{r=0}^{N-1} f_r \exp\left[-\mathrm{i}\left(\frac{\omega}{T} r\Delta t\right)\right]\Delta t \tag{6.16}$$

where $\Delta t = T/N$ and $t = r\Delta t$ substituting $T = N\,\Delta t$ into (6.16) gives,

$$F_k = \frac{1}{N}\sum_{r=0}^{N-1} f_r \mathrm{e}^{-\mathrm{i}\omega r/N} \qquad k = 0, 1, 2, \ldots, N-1 \tag{6.17}$$

hence the inverse discrete Fourier transform is given by

$$f_r = \sum_{k=0}^{N-1} F_k \mathrm{e}^{\mathrm{i}\omega r/N} \qquad r = 0, 1, 2, \ldots, N-1 \tag{6.18}$$

By the use of the fast Fourier transform algorithm these calculations may be considerably simplified in the computer, making the process very efficient for evaluating integrals of the form given by (f) in Example 6.1.[8]

6.4 LAPLACE TRANSFORMS[10]

This is closely related to the Fourier transform but generally with some important differences. If $f(t)$ is a piecewise differentiable function with a finite number of finite discontinuities we may define the Laplace transform $F(\kappa)$ of $f(t)$ by

$$F(\kappa) = L[f(t)] = \int_0^\infty f(t)\mathrm{e}^{-\kappa t}\,\mathrm{d}t \tag{6.19}$$

$f(t)$ is conventionally regarded as zero for $t < 0$. We shall indicate that we are working with the Laplace rather than the Fourier transform by using κ as the transform variable in our transformed function F. Note that κ may be complex and $F(\kappa)$ is defined in the first instance for $\mathrm{Re}(\kappa) > \gamma$ where γ is some constant defined for convergence of the integral (6.19) and Re means we take the real part. By integration by parts we can show that

$$L[\partial f/\partial t] = \kappa F(\kappa) - f(0) \tag{6.20}$$

where $f(0)$ is $f(t)$ evaluated at $t = 0$.

Similarly,

$$L[\partial^2 f/\partial t^2] = \kappa^2 F(\kappa) - \kappa f(0) - f'(0) \tag{6.21}$$

where $f'(0)$ is $\partial f/\partial t$ evaluated at $t = 0$.

Using (6.20) we may find the Laplace transform of an integral of $f(t)$ such that,

$$L\left[\int_0^t f(\tau)\,\mathrm{d}\tau\right] = \frac{F(\kappa)}{\kappa} \tag{6.22}$$

We shall also use the 'shifting theorem' which states that

$$L[\mathrm{e}^{at} f(t)] = \int_0^\infty f(t)\,\mathrm{e}^{(a-\kappa)t}\,\mathrm{d}t = F(\kappa - a) \tag{6.23}$$

Hence we have the important result that

$$L[\mathrm{e}^{at}] = \frac{1}{\kappa - a} \quad \text{and} \quad L[t\mathrm{e}^{at}] = \frac{1}{(\kappa - a)^2} \qquad (6.24), (6.25)$$

This result will be important when we carry out the numerical inversion of the Laplace transform.

As an analogy with (6.13) one can also define a Laplace convolution by,

$$f * g = \int_0^t f(t - \tau)\,g(\tau)\,\mathrm{d}\tau \tag{6.26}$$

Thus,

$$L[f * g] = F(\kappa)\,G(\kappa) \tag{6.27}$$

where,

$$F(\kappa) = L[f(t)] \quad G(\kappa) = L[g(t)] \tag{6.28}$$

Before considering the problems of the inversion of the Laplace transform, we shall consider its use in the solution of the simple second-order differential equation with constant coefficients discussed in Example 6.1.

Example 6.2

Consider the equation

$$m\frac{\mathrm{d}^2u}{\mathrm{d}t^2}+c\frac{\mathrm{d}u}{\mathrm{d}t}+ku=f(t) \tag{a}$$

now with the initial conditions $u(0)=u_0$, $\dot{u}(0)=\dot{u}_0$.

If we now take the Laplace transform of this equation we obtain,

$$m\kappa^2 U(\kappa)-m\kappa u_0-m\dot{u}_0+c\kappa U(\kappa)-cu_0+k\,U(\kappa)=F(\kappa) \tag{b}$$

Hence,

$$U(\kappa)=\frac{F(\kappa)+(m\kappa+c)u_0+m\dot{u}_0}{m\kappa^2+c\kappa+k} \tag{c}$$

$u(t)$ may then be obtained by using tables of transforms, by analytical evaluation of the Bromwich inversion integral or by a numerical inversion process.

Note that the initial values are taken into account in equation (c), which gives the 'transient' part of the solution.

THE BROMWICH INTEGRAL

We have,

$$F(\kappa)=\int_0^\infty \mathrm{e}^{-\kappa t}f(t)\,\mathrm{d}t \tag{6.29}$$

where $\mathrm{Re}(\kappa)>\gamma$ for convergence.

Then the use of Planchard's theorem gives[9]

$$f(t)=\int_{\gamma_1-\mathrm{i}\infty}^{\gamma_1+\mathrm{i}\infty}\mathrm{e}^{\kappa t}F(\kappa)\,\mathrm{d}\kappa \tag{6.30}$$

This is the Bromwich integral (see Figure 6.2). γ_1 is chosen to be such that equation (6.30) gives $f(t)\equiv 0$ for $t<0$. In fact γ_1 is chosen such that all the singularities of $F(\kappa)$ lie to the left of the contour from $\gamma_1-\mathrm{i}\infty$ to $\gamma_1+\mathrm{i}\infty$. By a minor distortion of the contour we may write (for $t>0$),

$$f(t)=\sum_j e^{\kappa_j t}F_j(\kappa) \tag{6.31}$$

where $F_j(\kappa)$ is the 'residue' at the jth singularity of $F(\kappa)$ which occurs at $\kappa = \kappa_j$ say.

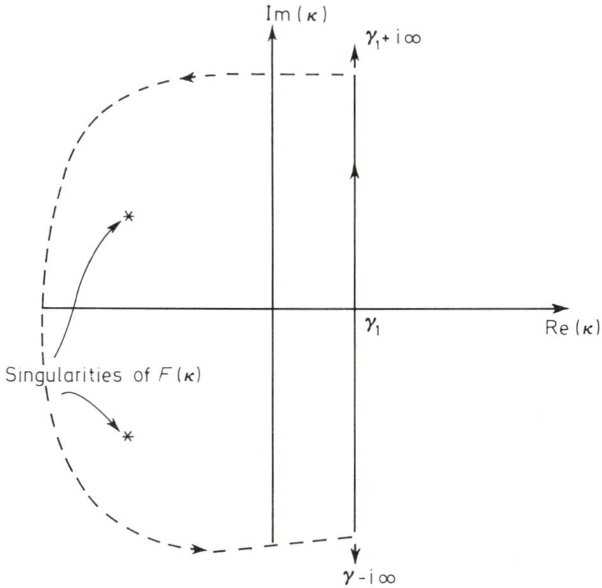

Figure 6.2 Contour in the complex κ plane for the Bromwich integral

The residue is the coefficient of the term involving $1/(\kappa - \kappa_j)$ in the Laurent expansion of $F(\kappa)$ about κ. (See reference 11 for a full explanation of this result.) The Laurent expansion in complex variable theory is similar to the Taylor expansion for functions of a real variable.

NUMERICAL INVERSION OF THE LAPLACE INTEGRAL

In our applications it may occur that the function $F(\kappa)$ is not known in closed form so that we shall only have available a number of sample points of F at the values of κ for which we have performed our boundary element calculations. In this case a numerical inversion technique will be needed. Following Schapery[11] we propose an exponential series to describe the solution of our problem variable $u(t)$ accurately. This involves finding a sufficient number of solutions for differing κ, so that an accurate curve fit to the equation

$$U(\kappa) = \sum_{j=1}^{n} \frac{S_j}{\kappa - \alpha_j} \tag{6.32}$$

can be obtained, i.e. to find the constants α_j and S_j ($j = 1, 2, \ldots, n$). Having obtained this curve, direct inversion will yield a solution of the form,

$$u(t) = \sum_{j=1}^{n} S_j \, e^{\alpha_j t} \tag{6.33}$$

We shall now discuss the application of the Laplace transform method to the problem of transient elastodynamics.

6.5 TRANSIENT ELASTODYNAMICS

Cruse[1] has solved this problem starting with the field equations based on the small displacement theory for a linear, isotropic, homogeneous, elastic medium (see Chapter 5).

We denote the displacement vector (in two or three dimensions) associated with each point $\vec{x}$ in space by,

$$\vec{u}(\vec{x}, t) = (u_j(\vec{x}, t)) \tag{6.34}$$

Then the equations of motion of the linear, isotropic, elastic body are,

$$(c_1^2 - c_2^2)u_{i,\,ij} + c_2^2\, u_{j,\,ii} + b_j = \partial^2 u_j/\partial t^2 \tag{6.35}$$

where $(\;)_{,j}$ means differentiation with respect to x_j.

c_1 is the velocity of dilatational waves where

$$c_1^2 = \frac{\lambda + 2\mu}{\rho} \tag{6.36}$$

with λ and μ the Lamé constants given by,

$$\lambda = \frac{E\nu}{(1-\nu)(1-2\nu)}, \quad \mu = \frac{E}{2(1+\nu)} \tag{6.37}$$

c_2 is the velocity of distortional waves where,

$$c_2^2 = \mu/\rho \tag{6.38}$$

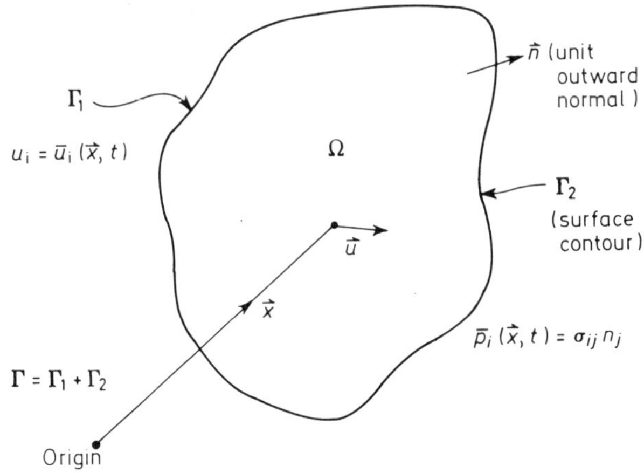

Figure 6.3 Problem region for elastodynamics problems

We have also the boundary conditions (see Figure 6.3)

$$u_i = \bar{u}_i(\vec{x}, t) \qquad \text{on } \Gamma_1 \tag{6.39}$$

$$p_i = \sigma_{ij} n_j = \bar{p}_i(\vec{x}, t) \qquad \text{on } \Gamma_2 \tag{6.40}$$

and initial conditions,

$$\left.\begin{aligned} u_i(\vec{x}, 0^+) &= u_{i0}(\vec{x}) \\ \frac{\partial}{\partial t}(u_i(\vec{x}, 0^+)) &= \dot{u}_{i0}(\vec{x}) \end{aligned}\right\} \quad \text{in } \Omega \text{ and } \Gamma \tag{6.41}$$

We also have the constitutive equation,

$$\sigma_{ij} = \rho[(c_1^2 - 2c_2^2) u_{k,k} \delta_{ij} + c_2^2(u_{i,j} + u_{j,i})] \tag{6.42}$$

We can now take the Laplace transform of (6.35) assuming sufficient continuity of the $u_{i,ij}$ terms. This gives,

$$(c_1^2 + c_2^2) U_{i,ij} + c_2^2 U_{j,ii} + B_j - \kappa^2 U_j + \dot{u}_{j0} + \kappa u_{j0} = 0 \tag{6.43}$$

where

$$U_i(\vec{x}, \kappa) = L(u_i(\vec{x}, t)) \tag{6.44}$$

and

$$B_i(\vec{x}, \kappa) = L(b_i(\vec{x}, t)) \tag{6.45}$$

The boundary conditions (6.39) and (6.40) may also be transformed in the usual way to give,

$$U_i(\vec{x}, \kappa) = \bar{U}_i(\vec{x}, \kappa) \qquad \text{on } \Gamma_1 \tag{6.46}$$

$$P_i(\vec{x}, \kappa) = \bar{P}_i(\vec{x}, \kappa) \qquad \text{on } \Gamma_2 \tag{6.47}$$

If we group the body force terms together we can write,

$$Q_j = \rho(B_j + \dot{u}_{j0} + \kappa u_{j0}) \tag{6.48}$$

and hence (6.43) becomes,

$$(c_1^2 - c_2^2)U_{i,ij} + c_2^2 U_{j,ii} - \kappa^2 U_j = -Q_j/\rho \tag{6.49}$$

Following Doyle[12] and utilising the Galerkin vector we may deduce the fundamental influence tensors for this equation as,

$$U_{ij}^* = \frac{1}{\alpha\pi\rho c^2}(\psi\delta_{ij} - \chi r_{,i} r_{,j}) \tag{6.50}$$

($\alpha = 2$ for two dimensions and $\alpha = 4$ for three).

For two dimensions ψ and χ are given by

$$\begin{aligned} \psi &= K_0\left(\frac{\kappa r}{c_2}\right) \frac{c_2}{\kappa r}\left[K_1\left(\frac{\kappa r}{c_2}\right) - \frac{c_2}{c_1} K_1\left(\frac{\kappa r}{c_1}\right)\right] \\ \chi &= K_2\left(\frac{\kappa r}{c_2}\right) - \frac{c_2^2}{c_1^2} K_2\left(\frac{\kappa r}{c_1}\right) \end{aligned} \tag{6.51}$$

K_0, K_1 and K_2 are modified Bessel functions of the third kind of orders zero, one and two respectively.[13]

For three dimensions,

$$\begin{aligned} \psi &= \frac{e^{-\kappa r/c_2}}{r} + \left(\frac{c_2^2}{\kappa^2 r^2} + \frac{c_2}{\kappa r}\right)\frac{e^{-\kappa r/c_2}}{r} - \frac{c_2^2}{c_1^2}\left(\frac{c_1^2}{\kappa^2 r^2} + \frac{c_1}{\kappa r}\right)\frac{e^{-\kappa r/c_1}}{r} \\ \chi &= \left(\frac{3c_2^2}{\kappa^2 r^2} + \frac{3c_2}{\kappa r} + 1\right)\frac{e^{-\kappa r/c_2}}{r} - \left(\frac{c_2^2}{c_1^2}\right)\left(\frac{3c_1^2}{\kappa^2 r^2} + \frac{3c_1}{\kappa r} + 1\right)\frac{e^{-\kappa r/c_1}}{r} \end{aligned} \tag{6.52}$$

The fundamental traction tensor components P_{ij}^* may be obtained using (6.40) and (6.42).

Using weighted residuals as previously shown we may obtain the transformed dynamic equivalent of the boundary integrals of Chapter 5. Note that U and P are the transformed displacements and tractions. For an internal point,

$$U_j + \int_\Gamma U_i P_{ji}^* \,\mathrm{d}\Gamma = \int_\Gamma P_i U_{ji}^* \,\mathrm{d}\Gamma + \int_\Omega Q_i U_{ji}^* \,\mathrm{d}\Omega \tag{6.53}$$

Using (6.42) a similar relationship could be obtained for the transformed stress tensor $\mathscr{L}(\sigma_{ij})$.

If we now take the point under consideration to the boundary Γ, we obtain the starting integral expression for the boundary element discretisation,

$$cU_j + \int_\Gamma U_i P_{ji}^* \,\mathrm{d}\Gamma = \int_\Gamma P_i U_{ji}^* \,\mathrm{d}\Gamma + \int_\Omega Q_i U_{ji}^* \,\mathrm{d}\Omega \tag{6.54}$$

For a smooth boundary the constant $c = \frac{1}{2}$.

Once (6.54) has been solved for a sufficient number of values of our transform parameter κ a numerical inversion process must be performed on $U_j(\vec{x}, \kappa)$ to obtain the time-dependent displacements $u_j(\vec{x}, t)$. The procedure used by Cruse[1] for this inversion is given by Bellman.[14]

6.6 STEADY STATE ELASTODYNAMICS

We shall now examine the problem considered above from the point of view of finding the steady state solution to the elastodynamics problem. We saw in Section 6.3 that the application of the Fourier transform to a differential equation resulted in a solution which was independent of the initial conditions of the problem, the particular integral.

If we start with the equation of motion of an isotropic elastic body, equation (6.35),

$$(c_1^2 - c_2^2)u_{i,ij} + c_2^2 u_{j,ii} + b_j = \partial^2 u_j / \partial t^2 \tag{6.55}$$

with boundary conditions

$$u_i = \bar{u}_i(\vec{x}, t) \qquad \text{on } \Gamma_1 \tag{6.56}$$

$$p_i = \sigma_{ij} n_j = \bar{p}_i(\vec{x}, t) \qquad \text{on } \Gamma_2 \tag{6.57}$$

If we now take the Fourier transform of these equations we obtain,

$$(c_1^2 - c_2^2)U_{i,\,ij} + c_2^2 U_{j,\,ii} + B_j + \omega^2 U_j = 0 \tag{6.58}$$

$$U_i(\vec{x}, \omega) = \bar{U}_i(\vec{x}, \omega) \qquad \text{on } \Gamma_1 \tag{6.59}$$

$$P_i(\vec{x}, \omega) = \bar{P}_i(\vec{x}, \omega) \qquad \text{on } \Gamma_2 \tag{6.60}$$

where now,

$$\begin{aligned} U_i(\vec{x}, \omega) &= \mathscr{F}[u_i(\vec{x}, t)] \\ B_i(\vec{x}, \omega) &= \mathscr{F}[b_i(\vec{x}, t)] \end{aligned} \tag{6.61}$$

Note that now the initial conditions do not enter into the formulation.

By making the substitution $\kappa = -\mathrm{i}\omega$ in equations (6.50) to (6.52) and choosing the correct Bessel functions and exponentials to give outgoing waves from our source points we obtain the following fundamental influence tensor components for this equation:

$$U_{ij}^{*} = \frac{1}{\alpha\pi\rho c_2^2}\left(\psi\delta_{ij} - \chi r_{,i} r_{,j}\right) \tag{6.62}$$

($\alpha = 2$ for two dimensions and $\alpha = 4$ for three).

For two dimensions,

$$\begin{aligned} \psi &= -\frac{\mathrm{i}\pi}{2} H_0^{(2)}\left(\frac{\omega r}{c_2}\right) + \frac{\mathrm{i}c_2\pi}{\omega r^2}\left[H_1^{(2)}\left(\frac{\omega r}{c_2}\right) - \frac{c_2}{c_1} H_1^{(2)}\left(\frac{\omega r}{c_1}\right)\right] \\ \chi &= \frac{\mathrm{i}\pi}{2}\left[H_2^{(2)}\left(\frac{\omega r}{c_2}\right) - \frac{c_2^2}{c_1^2} H_2^{(2)}\left(\frac{\omega r}{c_1}\right)\right] \end{aligned} \tag{6.63}$$

We have used the relationship[14]

$$K_n(-\mathrm{i}\omega) = -\tfrac{1}{2}\mathrm{i}\pi\,(\mathrm{i})^n H_n^2(\omega) \tag{6.64}$$

where $H_n^{(2)}$ is the Hankel function of the second kind.

For three dimensions

$$\begin{aligned} \psi &= \frac{\mathrm{e}^{\mathrm{i}\omega r/c_2}}{r} + \left(-\frac{c_2^2}{\omega^2 r^2} + \frac{\mathrm{i}c_2}{\omega r}\right)\frac{\mathrm{e}^{\mathrm{i}\omega r/c_2}}{r} + \frac{c_2^2}{c_1}\left(\frac{c_1^2}{\omega^2 r^2} - \mathrm{i}\frac{c_1}{\omega r}\right)\frac{\mathrm{e}^{\mathrm{i}\omega r/c_1}}{r} \\ \chi &= -\left(\frac{3c_2^2}{\omega^2 r^2} - \frac{3c_2}{\omega r} + 1\right)\frac{\mathrm{e}^{\mathrm{i}\omega r/c_2}}{r} + \left(\frac{c_2^2}{c_1^2}\right)\left(\frac{3c_1^2}{\omega^2 r^2} - \frac{3c_1}{\omega r} + 1\right)\frac{\mathrm{e}^{\mathrm{i}\omega r/c_1}}{r} \end{aligned} \tag{6.65}$$

Note that these expressions are complex.

In the same way as before we can obtain the integral equation in terms of our transformed variables.

$$cU_j + \int_\Gamma U_i P^*_{ji}\,\mathrm{d}\Gamma = \int_\Gamma P_i U^*_{ji}\,\mathrm{d}\Gamma + \int_\Omega \rho B_i U^*_{ji}\,\mathrm{d}\Omega \qquad (6.66)$$

Note that the integral over Ω only involves the transform of the body force appearing in the original equation (6.55). As before we solve (6.66) for a sufficient number of values of ω and numerically invert the transformed variable $U_j(\vec{x}, \omega)$ to obtain the time-dependent displacements $u_j(\vec{x}, t)$. A direct consideration of the magnitudes of $U_j(\vec{x}, \omega)$ for different ω will give the natural modes of vibration of the elastic body.

6.7 VISCOELASTIC PROBLEMS

Viscoelastic laws can be applied to materials to represent creep if the rate at which the creep strain develops depends not only on the current state of stress and strain but in general, on the full history of the applied loading.

For a linearly viscoelastic material the total stress at time t due to the initial strain and the incremental changes in strain is[15]

$$\begin{aligned}\sigma_{ij}(t) = {} & \lambda(t)\,\delta_{ij} u^0_{k,k} + \mu(t)\,(u^0_{j,i} + u^0_{i,j}) + \\ & + \int_0^t \lambda(t-\tau)\delta_{ij}\,\dot{u}_{k,k}(\tau)\,\mathrm{d}\tau + \\ & + \int_0^t \mu(t-\tau)\,(\dot{u}_{j,i}(\tau) + \dot{u}_{i,j}(\tau))\,\mathrm{d}\tau \end{aligned} \qquad (6.67)$$

where t is the time since loading, u^0 are the initial components of strain and λ, μ are the time-dependent Lamé parameters (they are now time dependent and can be interpreted as relaxation moduli).

Note that equation (6.67) can be written as a Laplace convolution, i.e.

$$\begin{aligned}\sigma_{ij}(t) = {} & \lambda\delta_{ij} u^0_{k,k} + \mu(u^0_{j,i} + u^0_{i,j}) + \\ & + \lambda_* \dot{u}_{k,k}\delta_{ij} + \mu_*(\dot{u}_{j,i} + \dot{u}_{i,j}) \end{aligned} \qquad (6.68)$$

Consider for simplicity only one component. We can do this by assuming that the bulk modulus $(E/3(1-2\nu))$ is constant or that the

Poisson ratio is constant. Let us use the second assumption with $\nu = 0$. This gives $\lambda = 0$ and reduces μ to $E/2$. Hence equation (6.68) becomes,

$$\sigma_{ij}(t) = \mu\gamma_{ij}^0 + \mu_*\gamma_{ij}^0 \tag{6.69}$$

where $\gamma_{ij} = (u_{j,i} + u_{i,j}) = \gamma_{ji}$.

In order to represent the time-dependent behaviour of the material McHenry[17] proposed the following equation to approximate the relaxation modulus:

$$\mu(t) = S_0 + \sum_{l=1}^{n} S_l(1 - e^{-t/\gamma l}) \tag{6.70}$$

This equation can be fitted to experimental data by reducing it to a set of linear equations in S_k and giving γ_l n values of time over the transition range of the experimental data.

Assuming $\gamma_{ij}^0 = 0$, equation (6.69) can be reduced to the following by taking the Laplace transform,

$$L(\sigma_{ij}) = \kappa L(\mu) L(\gamma_{ij}^0) \tag{6.71}$$

where from (6.70)

$$L(\mu(\kappa)) = \frac{1}{\kappa}\left[S_0 + \sum_{l=1}^{n}\left(\frac{S_l}{1+\gamma_l\kappa}\right)\right] \tag{6.72}$$

The stress–strain relationship of the viscoelastic material has thus been reduced to the normal static form ($\sigma_{ij} = \mu\gamma_{ij}$) in the transform plane (equation (6.71)). That is, we use a 'quasi-static' analysis.

To obtain a quasi-static analysis in boundary element terms we can set up an integral equation relationship for the displacements similar to equation (5.23), but using the constitutive relations (6.71) to construct the fundamental traction tensor components P_{ij}^*. κ will appear as a parameter of our problem and we must solve for a sufficient number of values in order that the inversion of our Laplace transforms can be accurately performed.

For the fully transient viscoelastic problem it is possible to follow Cruse[1] but instead of the constitutive equations (6.42) we have to use (6.68), which becomes equivalent to the normal static form under the Laplace transformation. The fundamental influence tensor U_{ij}^* would remain the same but the traction influence tensor P_{ij}^* would become altered by the new constitutive relations. The normal integral formulation (equation (6.54)) could then be used with these modifications and the inversions over the parameter κ performed in the usual way.

6.8 TIME INTEGRATION USING TIME STEPPING

If the time dependence in the governing equations cannot be removed using a transformation we have to integrate in time a hyperbolic or parabolic system.

The usual way of solving the problem is to use a time stepping type technique, where the problem is solved at each time interval. The evolution of the time-dependent terms is generally carried out in some finite difference way. Values at the interior of the region at any time may be calculated using the normal boundary element method.

Consider an equation of the type,

$$\mathscr{L}(u) = \partial u/\partial t \tag{6.73}$$

on the regions defined in Figure 6.4 (for simplicity we assume that u is a potential type function, although the same arguments apply to the

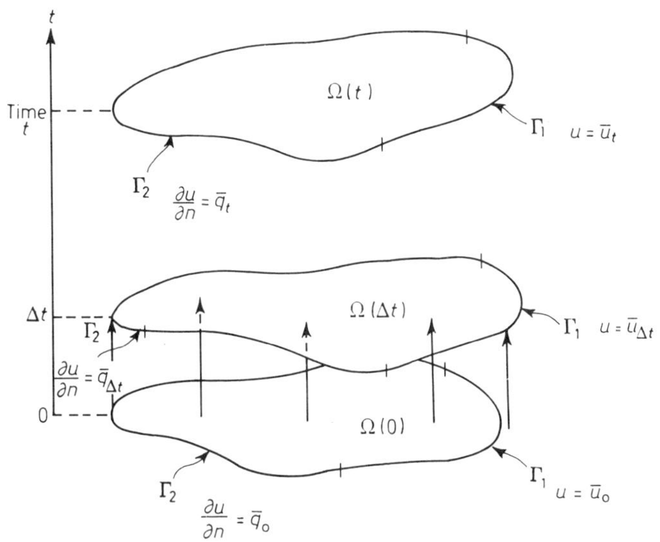

Figure 6.4 Time-dependent problem using time stepping technique

equations of elasticity). To solve the problem we require some initial values given over the whole of Ω at time $t = 0$, denoted by $\Omega(0)$. Say

$$u(\vec{x}, 0) = u_0 \tag{6.74}$$

We may also have time-dependent boundary conditions, i.e.

$$\begin{aligned} u &= \bar{u}_0 \quad \text{on } \Gamma_1 \quad \text{at time } 0 \\ u &= \bar{u}_t \quad \text{on } \Gamma_1 \quad \text{at time } t \end{aligned} \tag{6.75}$$

and

$$\begin{aligned} \partial u/\partial n &= \bar{q}_0 \quad \text{on } \Gamma_2 \quad \text{at time } 0 \\ \partial u/\partial n &= \bar{q}_t \quad \text{on } \Gamma_2 \quad \text{at time } t \end{aligned} \tag{6.76}$$

If we wish to find u at a later time Δt (we can call this value of u, $u_{\Delta t}$), where Δt is sufficiently small, we can write,

$$\mathscr{L}(u_{\Delta t}) \simeq \frac{u_{\Delta t} - u_0}{\Delta t} \tag{6.77}$$

or

$$\Delta t \, \mathscr{L}(u_{\Delta t}) - u_{\Delta t} = -u_0 \tag{6.78}$$

If we can find the fundamental solution to this equation, i.e. $u^*_{\Delta t}$, then we can use the boundary element method to solve for u at time Δt.

Example 6.3

Consider the equation of heat conduction in two dimensions, i.e.

$$\nabla^2 u = \partial u/\partial t \tag{a}$$

with

$$u = u_0 \quad \text{given at } t = 0 \quad \text{in } \Omega \tag{b}$$

$$u = \bar{u}_t \quad \text{on } \Gamma_1 \quad \text{at time } t \tag{c}$$

$$\partial u/\partial n = \bar{q}_t \quad \text{on } \Gamma_2 \quad \text{at time } t \tag{d}$$

Then equation (a) can be written as,

$$\nabla^2(u_{\Delta t}) - \frac{u_{\Delta t}}{\Delta t} = -\frac{u_0}{\Delta t} \tag{e}$$

The fundamental solution for (e) can be chosen to be (Helmholtz equation)

$$u^*_{\Delta t} = \frac{1}{2\pi} K_0 \left(\frac{r}{\sqrt{\Delta t}} \right) \tag{f}$$

which satisfies,

$$\nabla^2 u^*_{\Delta t} - \frac{1}{\Delta t} u^*_{\Delta t} = \delta_i \tag{g}$$

The weighted residual expression for $u_{\Delta t}$ becomes,

$$\int_\Omega (\nabla^2 u^*_{\Delta t} - \frac{1}{\Delta t} u^*_{\Delta t}) u_{\Delta t} \, d\Omega$$

$$= \int_\Gamma q^*_{\Delta t} u_{\Delta t} \, d\Gamma - \int_\Gamma u^*_{\Delta t} q_{\Delta t} \, d\Gamma + \frac{1}{\Delta t} \int_\Omega u_0 u^*_{\Delta t} d\Omega \tag{h}$$

This gives for our boundary values of $u_{\Delta t}$ the usual integral relations,

$$c_i u^i_{\Delta t} = \int_\Gamma q^*_{\Delta t} u_{\Delta t} \, d\Gamma - \int_\Gamma u^*_{\Delta t} q_{\Delta t} \, d\Gamma + \frac{1}{\Delta t} \int_\Omega u_0 u^*_{\Delta t} d\Omega \tag{i}$$

which is the starting point for our boundary element formulation.

The process may be extended to later times using the values found for u at Δt as our new initial values.

This method has the obvious disadvantage that errors may accumulate and u must be calculated at a sufficient number of internal points in order that they may be used as initial values for the following time step. This procedure may be very time consuming.

6.9 TIME-DEPENDENT FUNDAMENTAL SOLUTION

In this section we will consider another way of finding a time-dependent solution using boundary elements. Consider for instance,

$$\nabla^2 u = \frac{\partial u}{\partial t} \tag{6.79}$$

with boundary conditions of the same type as previously seen and some initial conditions. As the problem is now time dependent, the equation will also be integrated with relation to time. Weighting expression (6.79) we have,

$$\int_0^t \int_\Omega \left(\nabla^2 u - \frac{\partial u}{\partial t}\right) u^* \, \mathrm{d}\Omega \mathrm{d}\tau = \int_0^t \int_{\Gamma_2} (q - \bar{q})\, u^* \, \mathrm{d}\Gamma \mathrm{d}\tau - \\ - \int_0^t \int_{\Gamma_1} (u - \bar{u}) q^* \, \mathrm{d}\Gamma \mathrm{d}\tau \tag{6.80}$$

where $0 \leqslant \tau \leqslant t$.

Integrating by parts twice we find,

$$\int_0^t \int_\Omega \left(\nabla^2 u^* + \frac{\partial u^*}{\partial t}\right) u \mathrm{d}\Omega \mathrm{d}\tau - \left[\int_\Omega u u^* \, \mathrm{d}\Omega\right]_{\tau=0}^{\tau=t} + \\ + \int_0^t \int_\Gamma q u^* \, \mathrm{d}\Gamma \mathrm{d}\tau = \int_0^t \int_\Gamma u q^* \, \mathrm{d}\Gamma \mathrm{d}\tau \tag{6.81}$$

where the $\partial u^*/\partial t$ term was obtained integrating by parts with respect to time.

The corresponding fundamental solution for this equation is,[21]

$$u^* = \frac{1}{(4\pi\xi)^{d/2}} \exp\left(\frac{-r^2}{4\xi}\right) \tag{6.82}$$

where $\xi = t - \tau$ and d is the number of spatial dimensionality, i.e. $d = 3$ for three-dimensional problems, etc. the fundamental solution possesses the properties,

$$\nabla^2 u^* + \frac{\partial u^*}{\partial t} = 0 \qquad \text{in } \Omega \text{ for all } \tau \tag{6.83}$$

and for $\tau = t$:

$$\int_\Omega u^* \, \mathrm{d}\Omega = \begin{cases} 0 \text{ for } r \neq 0 \\ 1 \text{ for } r = 0 \end{cases}$$

Substituting this solution into equation (6.80), one obtains for a point 'i',

$$c_i u_i + \int_0^t \int_\Gamma u q^* \, \mathrm{d}\Gamma \mathrm{d}\tau = \int_0^t \int_\Gamma q u^* \, \mathrm{d}\Gamma \mathrm{d}\tau + \left[\int_\Omega u u^* \, \mathrm{d}\Omega\right]_{\tau=0} \tag{6.84}$$

The last term in the above formula corresponds to the initial conditions at $\tau = 0$. Since the fundamental solution itself is time dependent, one does not need to propose an iterative scheme to solve time-dependent problems as it is usually done in finite elements or finite differences.

When the surface is not smooth at the point 'i', one can still calculate the diagonal terms of $\mathbf{H}$ by the application of a uniform potential over the whole domain, which will give zero normal fluxes at the boundaries. Notice, however, that now equation (6.84) can be written as,

$$\mathbf{H}\,\mathbf{U} = \mathbf{G}\,\mathbf{Q} + \mathbf{P} \tag{6.85}$$

where the vector $\mathbf{P}$ depends on the potentials. Thus the coefficients on the diagonal of $\mathbf{H}$ will be given by,

$$H_{ii} = -\sum_{\substack{j=1 \\ (i \neq j)}}^{n} H_{ii} + P_i \tag{6.86}$$

The numerical procedure necessary to develop a computer program for time-dependent potential problems will now be shown, with reference to the two-dimensional unsteady heat conduction equation. In the absence of heat generators, this equation can be written as,

$$k\nabla^2 u = \frac{\partial u}{\partial t} \tag{6.87}$$

where u is temperature, t is time, $k = K/\rho c$, K is the thermal conductivity, ρ the mass density and c the specific heat. The fundamental solution of this equation is:[21]

$$u^* = \frac{1}{4\pi k(t-\tau)} \exp\left(-\frac{r^2}{4k(t-\tau)}\right) \tag{6.88}$$

One can divide the boundary of the domain under consideration into elements, as shown in Chapter 2. One also needs now to assume a variation on time for the functions u and q. As these functions vary much more slowly than u^* and q^*, it is a reasonable approximation to assume that they are constant over small intervals of time and perform the time integrations stepwise. Then, equation (6.84) becomes,

$$c_i u_i + k \int_\Gamma u \int_{t_1}^{t_2} q^* \, \mathrm{d}\tau \, \mathrm{d}\Gamma = k \int_\Gamma q \int_{t_1}^{t_2} u^* \, \mathrm{d}\tau \, \mathrm{d}\Gamma + \\ + \left[\int_\Omega u u^* \, \mathrm{d}\Omega \right]_{\tau = t_1} \tag{6.89}$$

Deriving equation (6.38) one finds,

$$q^* = \frac{\partial u^*}{\partial n} = -\frac{l}{8\pi k^2 (t-\tau)^2} \exp\left(-\frac{r^2}{4k(t-\tau)} \right) \tag{6.90}$$

where l is the distance defined by the normal from the point 'i' under consideration to the boundary element tangent. So the inner integral on the left-hand side of (6.89) can be carried out as follows,

$$\begin{aligned} \int_{t_1}^{t_2} q^* \, \mathrm{d}\tau &= \frac{l}{2\pi k} \int_{t_1}^{t_2} -\frac{1}{4k(t_2-\tau)^2} \exp\left(-\frac{r^2}{4k(t_2-\tau)} \right) \mathrm{d}\tau \\ &= -\frac{l}{2\pi k r^2} \exp\left(-\frac{r^2}{4k(t_2-t_1)} \right) \end{aligned} \tag{6.91}$$

To evaluate the time integral on the right-hand side of (6.89) one needs first to make an appropriate change of variables. Calling,

$$x = \frac{r^2}{4k(t_2-\tau)} \tag{6.92}$$

the integral becomes,

$$\begin{aligned} \int_{t_1}^{t_2} u^* \, \mathrm{d}\tau &= \frac{1}{\pi r^2} \int_{t_1}^{t_2} \frac{r^2}{4k(t_2-\tau)} \exp\left(-\frac{r^2}{4k(t_2-\tau)} \right) \mathrm{d}\tau \\ &= \frac{1}{4\pi k} \int_a^\infty \frac{\mathrm{e}^{-x}}{x} \, \mathrm{d}x = \frac{1}{4\pi k} \, \mathrm{Ei}[a] \end{aligned} \tag{6.93}$$

where

$$a = \frac{r^2}{4k(t_2-t_1)} \tag{6.94}$$

and $\mathrm{Ei}[a]$ is the exponential–integral function, which can be evaluated by series, i.e.

$$\mathrm{Ei}[a] = -C - \ln[a] + \sum_{n=1}^{\infty} (-1)^{n-1} \frac{a^n}{n.n!} \tag{6.95}$$

C being the Euler's constant, $C = 0{\cdot}57721566\ldots$

Equation (6.89) now becomes,

$$c_i u_i - \frac{1}{2\pi}\int_\Gamma \frac{u}{r^2}\exp\left(-\frac{r^2}{4k(t_2-t_1)}\right)\mathrm{d}\Gamma$$

$$= \frac{1}{4\pi}\int_\Gamma q\,\mathrm{Ei}[a]\,\mathrm{d}\Gamma + \frac{1}{4\pi k(t_2-t_1)}\int_\Omega u\exp\left(-\frac{r^2}{4k(t_2-t_1)}\right)\mathrm{d}\Omega \tag{6.96}$$

The integrations on space are carried out numerically. In the examples shown in Brebbia and Wrobel[22] constant elements were used and the boundary integrals calculated applying a 6 point Gauss quadrature formula. The integration over the domain was performed by dividing it into triangular cells and using a quintic quadrature scheme. The temperature on each integration point on the cells was evaluated at the end of each time step in order to determine the initial conditions for the next time step.

This new approach is of considerable interest as it eliminates the need of integrating step by step on time using finite differences or similar descretisations. In practical applications, however, small time intervals are recommended, but as shown in Brebbia and Wrobel[22], they can be very large without affecting the accuracy of the solution.

6.10 NON-LINEAR PROBLEMS

We have already discussed methods of dealing with materials which present hereditary constitutive laws of the linear viscoelastic type. We have also indicated how regions of a material, each with different properties, may be joined together. However, we have not yet discussed the problems for which the basic governing equations are themselves non-linear.

The boundary element method may seem at first sight to be inapplicable to non-linear problems as the source type solutions depend on superposition for their usefulness. We shall however indicate an iterative method which may supply a solution to some fairly well behaved weakly non-linear problems.

The non-linear terms can be treated as forcing functions, i.e. we transfer these terms to the right-hand side and treat them as known functions.

Consider the equation,

$$\mathscr{L}(u) + \mathscr{N}(u) = \mathscr{P} \tag{6.97}$$

where here $\mathscr{L}$ is a linear operator, $\mathscr{N}$ is a non-linear operator and $\mathscr{P}$ is some known forcing function.

The iterative procedure will consist of writing (6.97) as,

$$\mathscr{L}(u) = \mathscr{P} - \mathscr{N}(u) \tag{6.98}$$

making some realistic guess for the initial form of u, u_0 (or in some cases taking $u_0 \equiv 0$), and putting this in the non-linear terms on the right-hand side, i.e.

$$\mathscr{L}(u_1) = \mathscr{P} - \mathscr{N}(u_0) \tag{6.99}$$

If (6.99) is then solved for u_1 in the usual way we may repeat the process until some criterion for convergence is satisfied. The fundamental solution needed for our boundary element formulation when $\mathscr{L}$ is self-adjoint is defined by,

$$\mathscr{L}(u^*) = \delta_i \tag{6.100}$$

The resulting matrices for the boundary element formulation will be

$$\mathbf{HU} = \mathbf{GQ} + \mathbf{P} + \mathbf{N} \tag{6.101}$$

where **P** represents the body force vector and **N** is the vector obtained from the non-linear terms.

One important feature of this approach is that the influence matrices **H** and **G** remain the same for each iteration. We only need to recalculate the vector **N** and find the new solution vector **U** and **Q** for each iteration.

Example 6.4

Consider the equation,

$$\nabla^2 u + \left(\frac{\partial u}{\partial x}\right)^2 = 0 \quad \text{in } \Omega \tag{a}$$

and the fundamental solution given by,

$$\nabla^2 u^* = \delta_i \tag{b}$$

Integrating by parts the weighted residual statement corresponding to (a) we find that

$$c_i u_i = -\int_\Gamma u^* q \,\mathrm{d}\Gamma + \int_\Gamma q^* u \,\mathrm{d}\Gamma - \int_\Omega u^* \left(\frac{\partial u}{\partial x}\right)^2 \mathrm{d}\Omega \qquad \text{(c)}$$

The last term involves the integration of our unknown function so we cannot apply the boundary element method directly to this equation. As our first approximation we choose u_0 = constant such that

$$-\int_\Omega u^* \left(\frac{\partial u_0}{\partial x}\right)^2 \mathrm{d}\Omega = 0 \qquad \text{(d)}$$

and hence our first iteration $u = u_1$ will be given by the equation,

$$c_i u_1^i = -\int_\Gamma u^* q_1 \,\mathrm{d}\Gamma + \int_\Gamma q^* u_1 \,\mathrm{d}\Gamma \qquad \text{(e)}$$

The following iteration $u = u_2$ will be given by

$$c_i u_2^i = -\int_\Gamma u^* q_2 \,\mathrm{d}\Gamma + \int_\Gamma q^* u_2 \,\mathrm{d}\Gamma - \int_\Omega u^* \left(\frac{\partial u_1}{\partial x}\right)^2 \mathrm{d}\Omega \qquad \text{(f)}$$

and so on. For convergence we may require that the difference between two successive values of u, or their norms, is small.

6.11 PLASTICITY AND SOIL MECHANICS PROBLEMS

One important application of the boundary element method to non-linear problems is in the field of elastoplasticity.

The governing differential equations for the region of 'secondary creep' in elastoplasticity can be written[17] in terms of the incremental equilibrium equation, i.e.

$$\frac{\partial \dot{\sigma}_{ij}}{\partial x_j} + \dot{b}_i = 0 \qquad (6.102)$$

where ($\dot{}$) indicates stress rates. The stress rates are given by

$$\dot{\sigma}_{ij} = \lambda \delta_{ij} \dot{u}^{\mathrm{e}}_{k,k} + \mu(\dot{u}^{\mathrm{e}}_{j,i} + \dot{u}^{\mathrm{e}}_{i,j}) \qquad (6.103)$$

Note that $u^{\mathrm{e}}_{j,i}$ are the elastic strain components. If the total strain components are due to a recoverable (elastic) part and an unrecoverable plastic part, we can write,

$$u_{i,j} = u^{e}_{i,j} + u^{p}_{i,j} \tag{6.104}$$

Substituting (6.104) into (6.103) and then (6.103) into (6.102) we obtain,

$$(\lambda+\mu)\dot{u}_{i,ij} + \mu\,\dot{u}_{j,ii} + \dot{b}_j - [(\lambda+\mu)\dot{u}^{p}_{i,ij} + \mu\dot{u}^{p}_{j,ii}] = 0 \tag{6.105}$$

This equation can be written as,

$$(\lambda+\mu)\dot{u}_{i,ij} + \mu\,\dot{u}_{j,ii} + \dot{f}_i = 0 \tag{6.106}$$

with,

$$\dot{f}_i = -(\lambda+\mu)\dot{u}^{p}_{i,ij} - \mu\,\dot{u}^{p}_{j,i} + \dot{b}_j \tag{6.107}$$

f_i can be thought of as a fictitious body force term, which contains the plastic component of the deformation field and the traction forces are also modified as fo'lows:

$$\dot{p}_j + \mu(\dot{u}^{p}_{j,k} + \dot{u}^{p}_{k,j})$$

The fundamental influence tensors are the same as for the elastostatics problem but our variables now only represent increments, in addition an iterative must be used to find the fictitious body force vector $\dot{f}_i$. The $\dot{f}_i$ will appear in our integral expressions as an integral over the volume of our problem regions. The method is discussed in detail in Lin.[17] It is interesting to note that it is very similar to the initial strain approach which has been used extensively in finite elements.

Note that in order to calculate the plastic strain we need to define the yield function for the material, which is a function of the stresses (σ_{ij}) and a hardening parameter. We consider that the material is elastic-perfectly plastic (i.e. no hardening parameter is needed). Then we have that the yield function can be written as

$$\phi(\sigma_{ij}) = c \tag{6.108}$$

The parameter c defines the initial plastic state. The plastic strains are,

$$\varepsilon^{p}_{ij} = \tfrac{1}{2}(\dot{u}^{p}_{i,j} + \dot{u}^{p}_{j,i}) \tag{6.109}$$

The hypothesis of normality of the plastic strain increment to the yield surface leads to,

$$\dot{\varepsilon}^{p}_{ij} = \dot{\kappa}\,\partial\phi/\partial\sigma_{ij} \tag{6.110}$$

$\dot{\kappa}$ is a material parameter to be determined. The total strains are the sum of the elastic and plastic components, i.e.

$$\dot{\varepsilon}_{ij} = \dot{\varepsilon}^{e}_{ij} + \dot{\varepsilon}^{p}_{ij} = \dot{\varepsilon}^{e}_{ij} + \alpha\dot{\kappa}\frac{\partial \phi}{\partial \sigma_{ij}} \tag{6.111}$$

where

$$\alpha = \begin{cases} 1 & \text{when } \phi \geqslant c \\ 0 & \text{when } \phi < c \end{cases}$$

Applying the chain rule to equation (6.108) gives,

$$\frac{\partial \phi}{\partial \sigma_{ij}}\dot{\sigma}_{ij} = 0 \tag{6.112}$$

Inverting equations (6.111) and (6.112) and assuming $\alpha = 1$ we can obtain the following stress relations:

$$\dot{\sigma}_{ij} = \dot{\sigma}^{e}_{ij} + \dot{\sigma}^{p}_{ij} \tag{6.113}$$

The value of $\dot{\kappa}$ is then known.

Note that the incremental equilibrium equations can also be written as functions of the plastic strain components as

$$\mu\dot{u}_{i,jj} + (\lambda + \mu)\dot{u}_{j,ij} + \dot{b}_j - (\delta_{ij}\lambda\dot{\varepsilon}^{p}_{kk,j} + 2\mu\dot{\varepsilon}^{p}_{ij}) = 0 \tag{6.114}$$

The incremental elastoplastic treatment described above may be approximated in some situations by an elastic analysis incorporating non-linear stress–strain relations, with plastic yield modelled by yield surfaces along which special yield elements are placed.

Wood[18] has obtained some impressive results on the settlement of structures by coupling a simple plate finite element representation of the foundation with a boundary element representing the soil. The available analytic solutions were used for the force displacement characteristics of a homogeneous elastic half space, to take account of the heterogeneity in the soil the *heterogeneous* stress–strain relations were used to obtain the surface displacements. This approach has been shown to be surprisingly accurate.

In the field of underground excavations the boundary element method has been adapted by Bray[19] and Hocking.[20] The non-linearities in the soil are treated with an incremental approach as described above and so the path dependence of the stresses in the surrounding medium are taken into account. In addition special 'joint' elements are formulated to accommodate slip in faults in the excavated medium. Bray uses fictitious forces to represent the release of stress whereas Hocking uses a distribution of quadrupoles. The actual boundary of the excavated volume is represented by imposing fictitious forces on its boundary in order to enforce the zero applied tractions there. The analysis is then capable of determining the stresses and displacements in the surrounding medium after the excavation has taken place. There is good agreement between the numerical and analytic solutions, for the stresses and induced displacements around thin slits in a triaxial stress field as well as for a circular tunnel between soft horizontal seams.

REFERENCES

1. Cruse, T. A. and Rizzo, F. J., 'A Direct Formulation and Numerical Solution of the General Transient Elastodynamic Problem, I', *J. Math. Anal. Applic.*, **22**, No. 1, 244–59 (1968)
2. Cruse, T. A., 'A Direct Formulation and Numerical Solution of the General Transient Elastodynamic Problem. II', *J. Math. Anal. Applic.*, **22**, No. 2, 341–55 (1968)
3. Rizzo, F. J. and Shippy, D. J., 'An Application of the Correspondence Principle of Linear Viscoelasticity Theory', *SIAM J. Appl. Math.*, **21**, No. 2, 321–30
4. Shippy, D. J., 'Application of the Boundary-Integral Equation Method to Transient Phenomena in Solids', *Boundary Integral Equation Method: Computational Applications in Applied Mechanics*, AMD-Vol. II, ASME, New York (1975)
5. Hutchinson, J. R., 'Determination of Membrane Vibrational Characteristics by the Boundary-Integral Equation Method', *Proc. 1st Int. Seminar on Recent Advances in Boundary Element Methods, Southampton University, July 1978*, Pentech Press (1978)
6. Banerjee, P. K. and Mustoe, G. G., 'The Boundary Element Method for Two-Dimensional Problems of Elastoplasticity', *Proc. 1st Int. Seminar on Recent Advances in Boundary Element Methods, Southampton University, July 1978*, Pentech Press (1978)
7. Jeffreys, H. and Jeffreys B. S., *Methods of Mathematical Physics*, Cambridge University Press (1946)
8. Newland, D. E., *An Introduction to Random Vibrations and Spectral Analysis*, Longman, London and New York (1975)
9. Smith, G., *Laplace Transform Theory*, Van Nostrand (1963)
10. Carrier, F., Krook, M. and Pearson, C. E., *Functions of a Complex Variable*, McGraw-Hill (1952)
11. Schapery, R. A., 'Approximate Methods of Transform Inversion in Viscoelastic Stress Analysis', *Proc. 4th US Ntl. Congr. on Applied Mechanics*, **2**, 1075 (1962)
12. Doyle, J. M., 'Integration of the Laplace Transformed Equations of Classical Elastokinetics', *J. Math. Anal. Applic.*, **13**, (1966)
13. Watson, G. N. *A Treatise on the Theory of Bessel Functions*, 2nd Edn, Cambridge University Press (1958)
14. Bellman, R. E., Kalaba, R. E. and Lockett, J., *Numerical Inversion of the Laplace Transform*, American Elsevier, New York (1966)
15. Adey, R. A. and Brebbia, C. A., 'Efficient Method for Solution of Viscoelastic Problems', *Proc. ASCE J. Engng Mech. Div.*, Dec. (1973)
16. McHenry, D., 'A New Aspect of Creep in Concrete and its Application to Design',

Proc. Am. Soc. for Testing and Materials, **43** (1943)
17. Lin, T. H., *Theory of Inelastic Structures*, Wiley (1968)
18. Wood, L. A., 'A Simple Boundary Element Approach to the Prediction of the Settlement of Structures', *Proc. 1st Int. Seminar on Recent Advances in Boundary Element Methods, Southampton University, July 1978*, Pentech Press (1978)
19. Bray, J. W., 'A Program for Two-dimensional Stress Analysis using the Boundary Element Method', *Rock Mechanics Progress Report No. 16*, Imperial College of Science and Technology, London (1976)
20. Hocking, G., 'Stress Analysis of Underground Excavations Incorporating Slip and Separation along Discontinuities', *Proc. 1st Int. Seminar on Recent Advances in Boundary Element Methods, Southampton University, July 1978*, Pentech Press (1978)
21. Carslaw, H. S. and Jaeger, J. C., *Conduction of Heat in Solids*, 2nd edn, Clarendon Press, Oxford (1959)
22. Brebbia, C. A. and Wrobel, L. C., 'Steady and Unsteady Potential Problems using the Boundary Element Method'. In *Recent Advances in Numerical Methods in Fluids*, Ed. C. Taylor, Swansea, (1979)

7 Combination of Regions

7.1 INTRODUCTION

Boundary solutions offered important advantages over the 'domain' type techniques. In particular, we can use a smaller number of unknowns to analyse the same problems, as unknowns are not needed inside the domain. This characteristic renders boundary solutions specially attractive for three-dimensional problems for which the ratio of the external surface to the volume is small, such as concrete nuclear pressure vessels, gravity dams, etc.

The idea of having the unknowns varying over part of the surface with known interpolation functions gives origin to the boundary element technique. The technique is similar to taking finite elements only on the surface of the body. Boundary elements allow us to describe better the geometry of the body under consideration and in addition, are easy to combine with finite elements.

One of the main disadvantages of the finite element method is its obvious inability to model domains extending to infinity. Boundary elements on the contrary can use fundamental solutions which naturally obey the conditions at infinity. It may be difficult however to find fundamental solutions for all cases but some approximation may equally well be applied. This idea has not yet been fully exploited in boundary solutions, although it has been used with a certain degree of success for domain techniques, i.e. the so-called 'infinite elements'.[1,2]

Its inherent ability to model domains extending to infinity has made the boundary element method popular for modelling problems such as those occurring with the Helmholtz type equation, i.e. problems such as diffraction, harbour resonance, etc. There are also problems for which approximations could be used to represent the boundary conditions with good results. As mentioned, the full potential of these approximations has not yet been fully investigated.

One of the more interesting features of the boundary element method is the simplicity of the input data required to run a homogeneous problem. Instead, in finite elements and other domain methods the amount of data needed is usually very large. This is an important practical point as many man hours are usually spent in preparing and checking finite element data.

The accuracy of boundary elements is generally superior to that of finite elements, except for a narrow boundary layer. Boundary elements (being mixed type formulation) also produce good results for the two types of variables under consideration (i.e. potentials and fluxes or displacements and stresses). The method is better suited to analyse problems with stress or flux concentrations for which finite elements is generally unsuited due to the interpolation functions used.

The main disadvantages of boundary elements are the difficulties encountered in non-homogeneous problems (i.e. finding the fundamental solutions and defining the interfaces) and the fact that all matrices are now fully populated. In this chapter we will describe how these disadvantages can be overcome by subdividing the body into regions. This subdivision is also needed for bodies with different dimensions in different directions (e.g. slender beams).

7.2 DIVISION INTO REGIONS

In many practical applications it is necessary to divide the body into several regions. This may be due to having a non-homogeneous body or because the dimensions of the body are not regular. Long bodies for instance may produce numerical accuracy problems as the influence of some nodes over others becomes very weak.

Consider for instance the case of a two-dimensional body for simplicity divided into two different regions (Figure 7.1). The interface between the two regions will be called Γ_{I} and two different fundamental solutions will apply over the parts 1 and 2 of the domain.

Let us analyse the two parts separately. On the first region we have

$$[\mathbf{H}^1 \quad \mathbf{H}_{\mathrm{I}}^1]\begin{Bmatrix}\mathbf{U}^1\\ \mathbf{U}_{\mathrm{I}}^1\end{Bmatrix} = [\mathbf{G}^1 \quad \mathbf{G}_{\mathrm{I}}^1]\begin{Bmatrix}\mathbf{P}^1\\ \mathbf{P}_{\mathrm{I}}^1\end{Bmatrix} \tag{7.1}$$

where $\mathbf{U}^1$ and $\mathbf{P}^1$ are the displacements and tractions on the external surface of region 1 and $\mathbf{U}_I^1$ and $\mathbf{P}_I^1$ the displacements and tractions on the interface Γ_I, considering that it belongs to region 1.

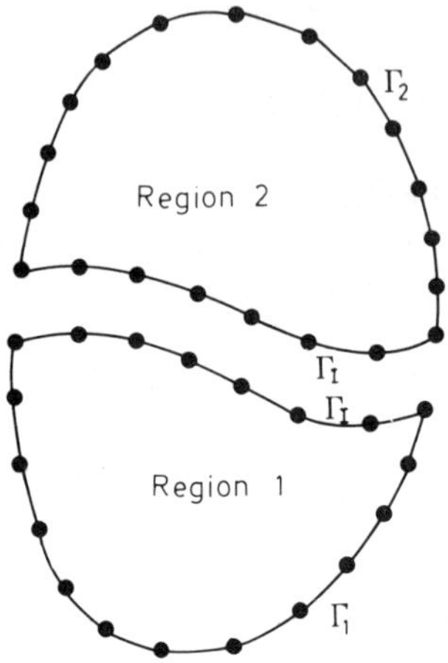

Figure 7.1 Body divided into two regions

For the second region we have,

$$[\mathbf{H}^2 \quad \mathbf{H}_I^2]\begin{Bmatrix}\mathbf{U}^2\\ \mathbf{U}_I^2\end{Bmatrix} = [\mathbf{G}^2 \quad \mathbf{G}_I^2]\begin{Bmatrix}\mathbf{P}^2\\ \mathbf{P}_I^2\end{Bmatrix} \tag{7.2}$$

where $\mathbf{U}^2$ and $\mathbf{P}^2$ are the displacements and tractions on the external surface of region 2 and $\mathbf{U}_I^2$ and $\mathbf{P}_I^2$ the displacements and tractions on the interface Γ_I, considering that it belongs to region 2.

If we call $\mathbf{U}_I$ and $\mathbf{P}_I$ the tractions and displacements on the interface Γ_I, we can write the following conditions:

$$\begin{aligned} &\text{(i) Compatibility} \quad \mathbf{U}_I = \mathbf{U}_I^1 = \mathbf{U}_I^2 \\ &\text{(ii) Equilibrium} \quad \mathbf{P}_I = \mathbf{P}_I^1 = -\mathbf{P}_I^2 \end{aligned} \tag{7.3}$$

Hence equations (7.1) and (7.2) can be rewritten as,

$$[\mathbf{H}^1 \quad \mathbf{H}_I^1 \quad -\mathbf{G}_I^1]\begin{Bmatrix}\mathbf{U}^1\\ \mathbf{U}_I\\ \mathbf{P}_I\end{Bmatrix} = \mathbf{G}^1\,\mathbf{P}^1 \tag{7.4}$$

$$[\mathbf{H}^2 \quad \mathbf{H}_I^2 \quad \mathbf{G}_I^2] \begin{Bmatrix} \mathbf{U}^2 \\ \mathbf{U}_I \\ \mathbf{P}_I \end{Bmatrix} = \mathbf{G}^2 \mathbf{P}^2 \tag{7.5}$$

Equations (7.4) and (7.5) can be written together as follows,

$$\underbrace{\begin{bmatrix} \mathbf{H}^1 & \mathbf{H}_I^1 & -\mathbf{G}_I^1 & \mathbf{0} \\ \mathbf{0} & \mathbf{H}_I^2 & \mathbf{G}_I^2 & \mathbf{H}^2 \end{bmatrix}}_{\text{interface}} \begin{Bmatrix} \mathbf{U}_I^1 \\ \mathbf{U}_I \\ \mathbf{P}_I \\ \mathbf{U}^2 \end{Bmatrix} = \begin{bmatrix} \mathbf{G}^1 & \mathbf{0} \\ \mathbf{0} & \mathbf{G}^2 \end{bmatrix} \begin{Bmatrix} \mathbf{P}^1 \\ \mathbf{P}^2 \end{Bmatrix} \tag{7.6}$$

The system needs still to be recorded according to the prescribed displacements and tractions.

This solution has the added advantage that the system of equations is now banded. Note that compatibility and equilibrium are explicitly satisfied between regions (in finite elements instead only compatibility is identically satisfied). For a body divided into three regions for instance the global matrix (7.6) after applying the boundary conditions and consequent reordering will look like that shown in Figure 7.2. The unknowns at the interface are the stresses and displacements.

Example 7.1 The Rolling Mill Cylinder[3]

The cylinder is one of a pair, supported by bearings at the ends and loaded (by the plate being rolled) along the central portion of its length. In Figure 7.3 the idealisation of the problem is shown; the bearings are represented by simple supports on a circle defined by the intersection of the surface of the bearing and a vertical plane bisecting that surface.

In the analysis presented here, the effect of temperature differentials is ignored. The load is in reality distributed over a finite area of the cylinder; in the analysis it is supposed that the load is distributed along the line through which the resultant passes. As may be seen from Figure 7.3, the resultant does not pass through the axis of the cylinder, so there is torsion as well as bending.

The object of the analysis is to determine the admissibility of vertical cracks located at the centre of the span. It is not necessary to

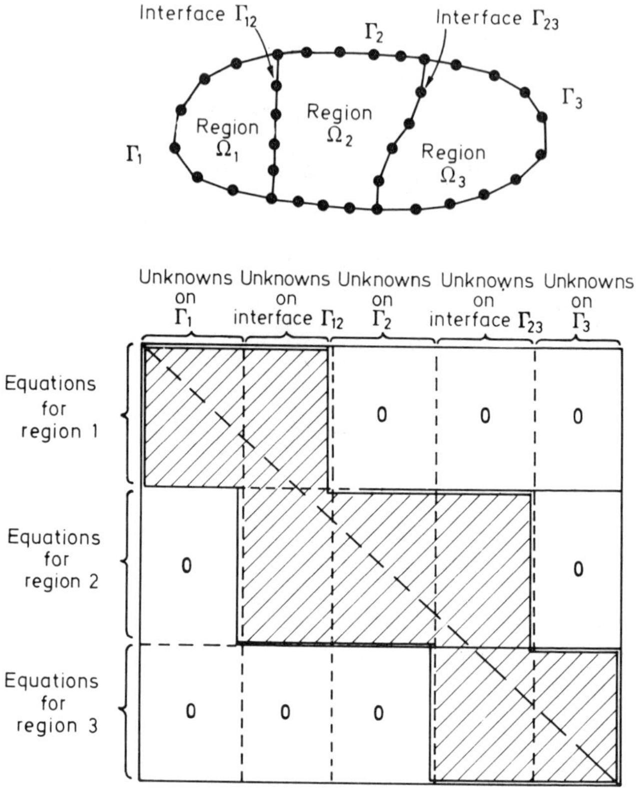

Figure 7.2 Banded matrix equations corresponding to a body divided into three regions

withdraw a cylinder from service immediately a crack appears; but if the cylinder fails whilst in operation, considerable damage is done to other components such as the bearings. For the purposes of analysis, the crack tip is supposed to be circular (Figure 7.4).

The modulus of elasticity was taken to be 210 000 N mm^{-2} and Poisson's ratio 0·3. The intensity of the line load was nearly constant, 8500 N mm^{-1}, and the length of the perpendicular from the axis of the cylinder to the resultant was 40 mm. There was therefore very little torsion and the critical case may be taken to be that in which the crack is diametrically opposed to the load point. The crack considered here was of depth 165 mm and radius 1100 mm. The cylinder was divided into 5 regions (Figure 7.5), the fifth containing the crack. The

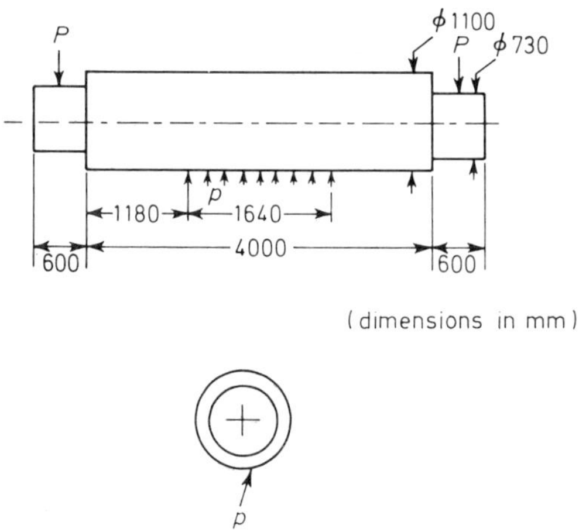

Figure 7.3 Idealisation of the rolling mill cylinder

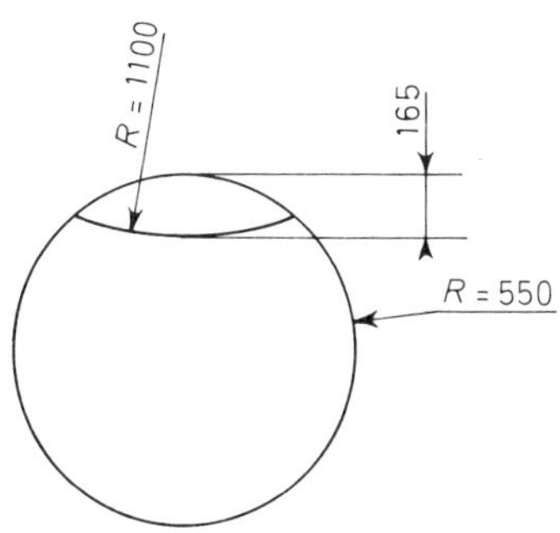

Figure 7.4 Dimensions of the crack considered

discretisation of the surface of each region is shown in Figures 7.6, 7.7 and 7.8.

Quadratic functional variation is chosen.

The program calculated the vertical displacement at mid-span to be 2·5 mm. This compared well with the deflection, 1·9 mm, obtained by the finite element method for the uncracked cylinder. The finite element analysis incidentally took 24 times longer to run on the computer.

Figure 7.9 shows the variation of calculated direct stress on the plane containing the crack as the crack tip is approached, along five lines in that plane.

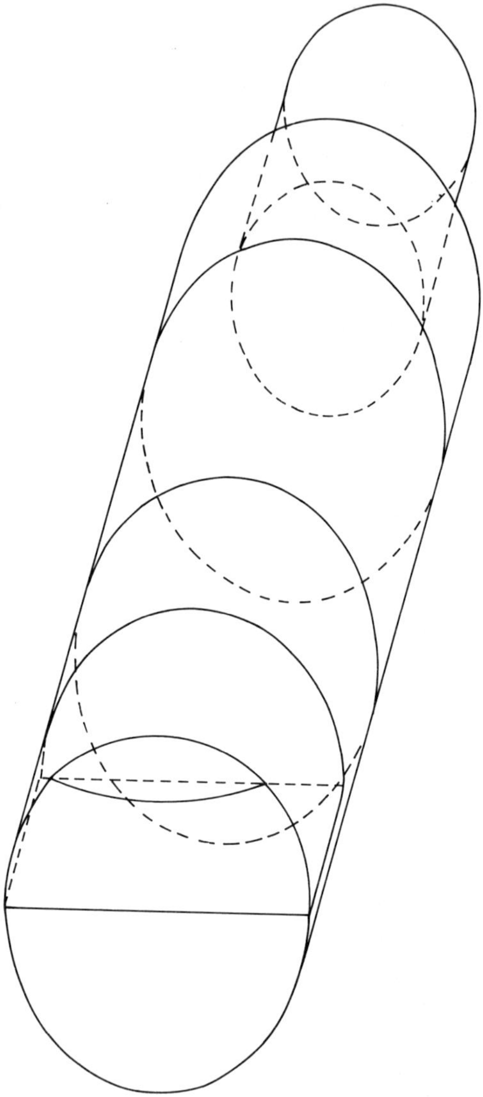

Figure 7.5 Division of the cylinder into regions

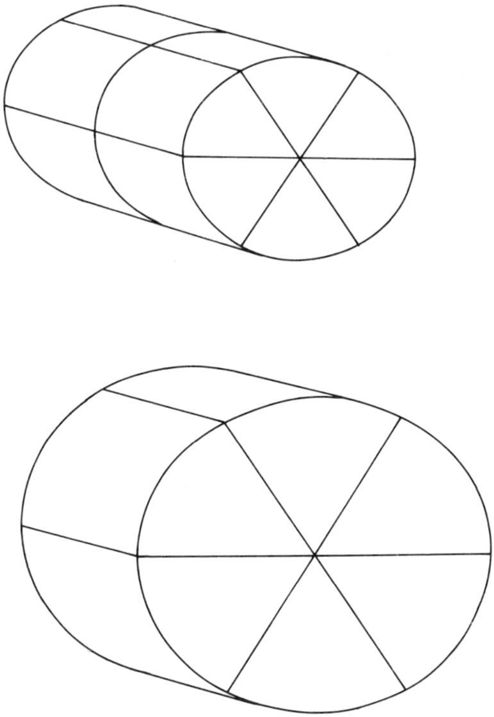

Figure 7.6 First and second regions

From these results the stress intensity factor was calculated, according to the linear theory of fracture mechanics to be

$$K_{\mathrm{I}} = \lim_{r \to 0} \sigma \sqrt{2\pi r}$$

The variation of stress intensity factor along the crack tip is shown in Figure 7.10.

7.3 APPROXIMATE BOUNDARY SOLUTIONS

We have already indicated that approximate solutions can be used to form boundary elements. These solutions are of practical interest

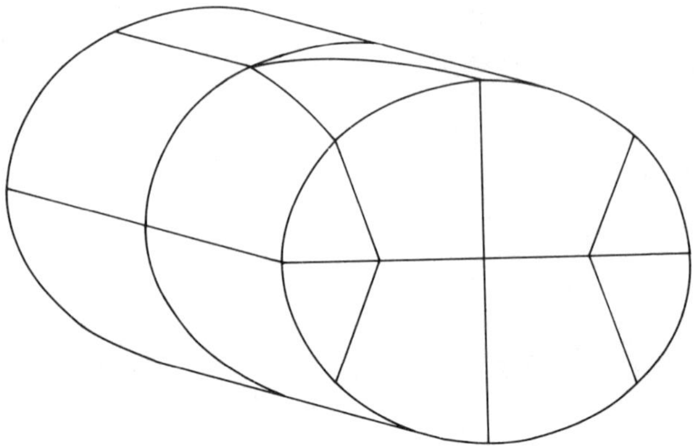

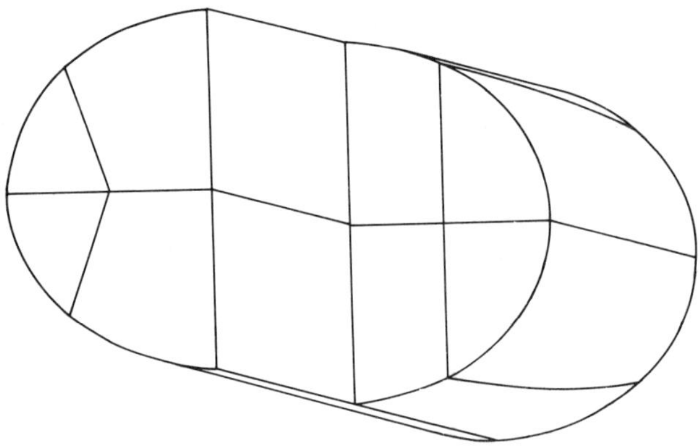

Figure 7.7 Third and fourth regions

when the fundamental solution is difficult to obtain or cumbersome to use.

To illustrate the technique consider the case of the Helmholtz equation, i.e.

$$\nabla^2 u + \kappa^2 u = 0 \tag{7.7}$$

κ is the wavenumber and u a potential. Let us consider that the domain can be divided into two regions, Ω_1 where a finite element discretisation is applied and Ω_2 which extends to infinity. The two regions are connected at the interface Γ_I (Figure 7.11).

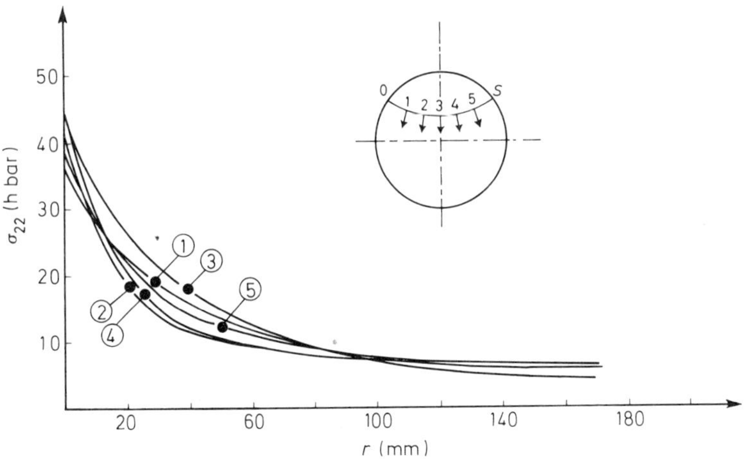

Figure 7.8 Fifth region

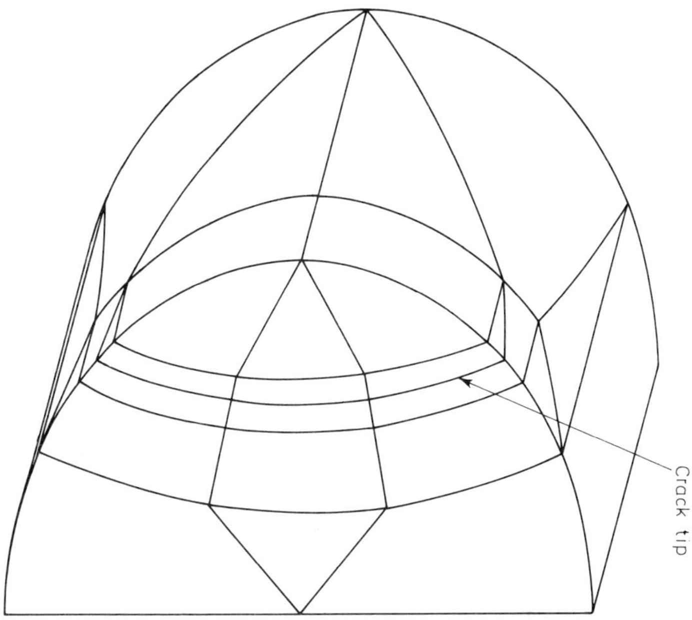

Figure 7.9 Variation of the calculated direct stress σ_{zz}

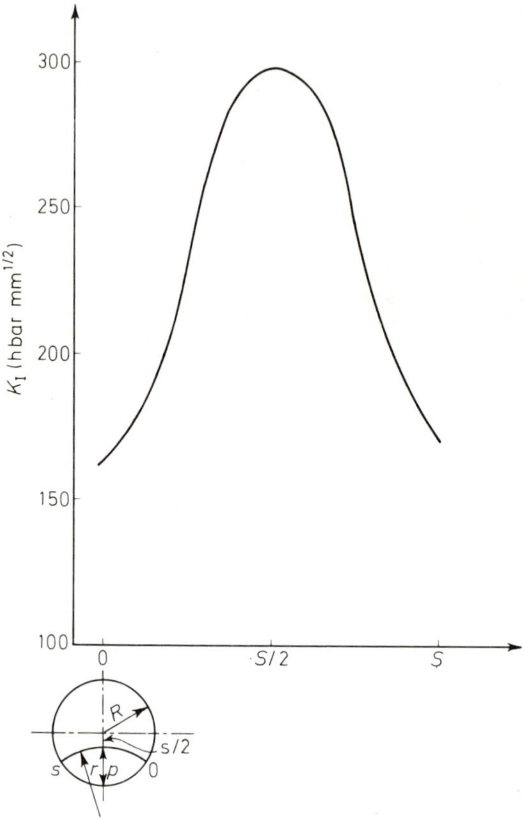

Figure 7.10 Variation of the stress intensity factor along the crack tip

Let us analyse the Ω_2 region considering that the elements on the interface are boundary elements. The fundamental solution of the Helmholtz equation for two dimensions is (assuming a time dependence $e^{i\omega t}$)

$$u^* = \frac{1}{4i} H_0^{(2)}(\kappa r) \tag{7.8}$$

where r is the distance from the point under consideration to any other point on Γ_I.

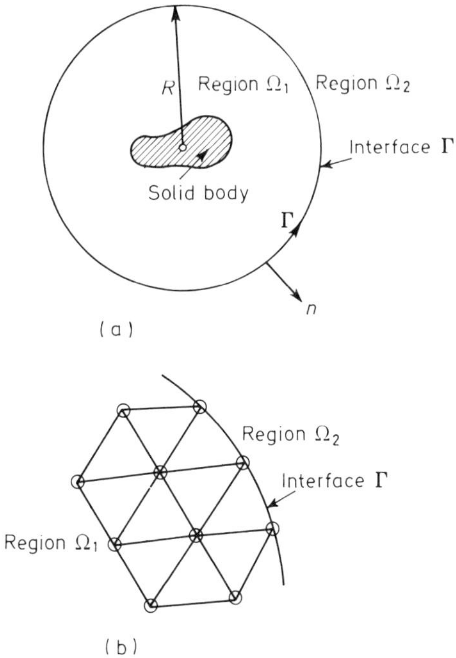

Figure 7.11 Finite and boundary elements domain: (a) *domain under consideration;* (b) *interface*

Weighting equation (7.7) by u^* and integrating by parts we obtain,

$$\int_{\Omega_2} (\nabla^2 u^* + \kappa^2 u^*)\, \hat{u}\, \mathrm{d}\Omega = \int_{\Gamma_\mathrm{I}} \frac{\partial u^*}{\partial n}\, \hat{u}\, \mathrm{d}\Gamma - \int_{\Gamma_\mathrm{I}} u^* \frac{\partial \hat{u}}{\partial n}\, \mathrm{d}\Gamma \qquad (7.9)$$

As the function u^* satisfies the Helmholtz equation with δ_i on the right-hand side we can write,

$$\int_{\Gamma} u^* \frac{\partial \hat{u}}{\partial n}\, \mathrm{d}\Gamma = \int \frac{\partial u^*}{\partial n}\, \hat{u}\, \mathrm{d}\Gamma \qquad (7.10)$$

for x_i in Ω_1. Once the form of the fundamental solution is substituted into (7.10) we can write,

$$\int_{\Gamma} H_0^{(2)}(\kappa r) \frac{\partial \hat{u}}{\partial n}\, \mathrm{d}\Gamma = \int_{\Gamma} \frac{\partial H_0^{(2)}(\kappa r)}{\partial n}\, \hat{u}\, \mathrm{d}\Gamma \qquad (7.11)$$

In this way a boundary element matrix equation could be obtained in the way previously shown. The equations can then be superimposed to the finite element solution for region Ω_1 and the problem solved along the lines described in Section 7.2. This approach has been used by Chen and Mei[5] and more recently by Shaw and Falby.[6] It has the disadvantage that the boundary values are all interrelated, resulting in non-banded equations, and that it may be cumbersome to work with Hankel's functions.

In the conventional boundary element approach the observation point is taken to be on the boundary. Each observation point gives us one equation connecting the unknowns $\partial u/\partial n$ and u, which are evaluated on the boundary. If, however, we take the observation points to be in the region Ω_2 far from our problem region we may make the following important simplification using the asymptotic forms for the Hankel function, hence

$$H_0^{(2)}(\kappa r) = \sqrt{\frac{2}{\pi\kappa r}}\exp[-\mathrm{i}(\kappa r - \pi/4)] \tag{7.12}$$

$$\frac{\partial H_0^{(2)}(\kappa r)}{\partial n} = -\kappa H_1^{(2)}(\kappa r) \simeq -\mathrm{i}\kappa\sqrt{\frac{2}{\pi\kappa r}}\exp[-\mathrm{i}(\kappa r - \pi/4)] \tag{7.13}$$

(note that now $n \simeq r$). Substituting these asymptotic forms into (7.11), taking Γ to be a large circle of radius R we obtain

$$\sqrt{\frac{2}{\pi\kappa}}\,\mathrm{e}^{\mathrm{i}\pi/4}\int_\Gamma \frac{1}{\sqrt{R}}\left(\frac{\partial u}{\partial R} + \mathrm{i}\kappa u \mathrm{e}^{-\mathrm{i}\kappa R}\right)\mathrm{d}\Gamma = 0 \tag{7.14}$$

If we now use the fact that R is constant on Γ we can write (7.14) as an integral over the angular coordinate θ as:

$$\int_0^{2\pi} \sqrt{R}\left(\frac{\partial u}{\partial R} + \mathrm{i}\kappa u\right)\mathrm{d}\theta = 0 \tag{7.15}$$

Note that this is just the form of the Sommerfeld radiation condition, which may be applied approximately on Γ as,

$$\frac{\partial u}{\partial r} + \mathrm{i}\kappa u = 0 \tag{7.16}$$

this condition has been successfully applied in many fluid mechanics problems. However, the present approach can be extended to develop other types of boundary elements, such as those to be used in soil mechanics problems.

Example 7.2

In order to determine how the application of the above radiation boundary condition affects the accuracy of the solution, the case of a single vertical column subject to an incident harmonic wave was studied.

The column was surrounded by a finite element mesh as shown in (Figure 7.12), and wave diffraction results were found for different wavelengths, with the application of condition (7.16) on the external boundary.

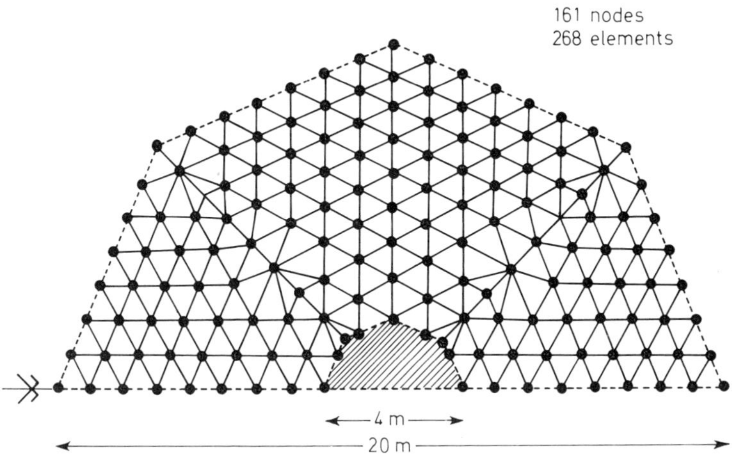

Figure 7.12 Finite element mesh

As a test to determine the adequacy of the radiation condition in representing a train of plane harmonic waves, the case with no solid cylinder was studied first. For long waves (λ, wavelength $\simeq$ 30 m) the results were very accurate, within 3 % of the exact solution. When the wavelength was reduced the errors tended to increase, which was to be expected as the element mesh became too coarse. It was found that for linear elements such as the ones used here, ideally 8 elements per wavelength should be used.

A further test was run to compare our results with those obtained by Mei[7] using a finite element mesh and the Hankel function formulation, i.e. the fundamental solution. The results are for an incident wave with wavelength $\lambda = 2\pi$ m and unit incident surface elevation for frequency $\omega = 3{\cdot}1321$. In Figure 7.13 the results of the exact solution, the finite element solution due to Mei and the solution obtained by the authors are compared. The present solution (7.9) compares favourably taking into consideration the coarseness of the

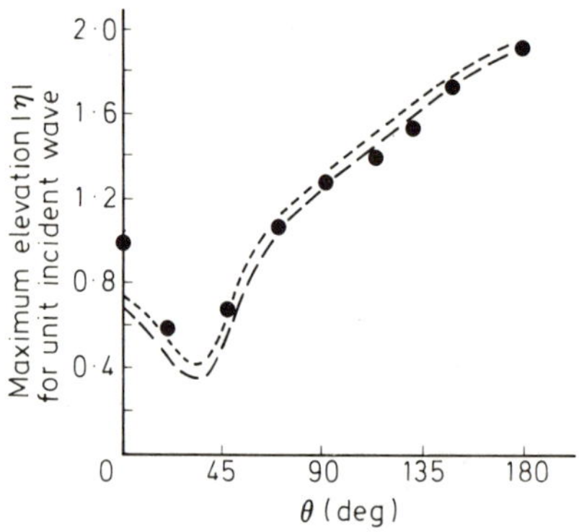

Figure 7.13 Results for maximum surface elevations round a circular cylinder: radius = 2 m, wave length = 2π m. – – –, from Mei[7]; ——— ——— ———, analytical; ●, authors' results (coarse mesh with linear interpolation)

mesh around the cyliner. Mei's solution for instance uses 18 elements round the cyliner and represents the geometry of the obstruction better.

Example 7.3 Harbour Resonance[8]

As another example of the application of this method we will consider the harbour resonance equation, i.e.

$$(u) = \frac{\partial}{\partial x}\left(h\frac{\partial u}{\partial x}\right) + \frac{\partial}{\partial y}\left(h\frac{\partial u}{\partial y}\right) + \kappa^2 h\,u = 0 \qquad \text{(a)}$$

with boundary condition

$$G(u) = h\frac{\partial u}{\partial n} = q \quad \text{on } \Gamma_2, \tag{b}$$

where u is the wave elevation referred to the still water level with in-plane coordinates x and y, h is the depth, q is a given value on Γ_2 boundaries and κ is the wavenumber.

We have the situation shown in Figure 7.14. We need to apply a boundary condition on the fictitious ocean boundary Γ_2 which will allow wave energy to pass across Γ_2 without distortion. In order to do this we need to split the wave elevation u into three parts:

(i) the incident field u_I;
(ii) the reflected field calculated from plane wave reflections by the coastline and called u_R (see Figure 7.14)

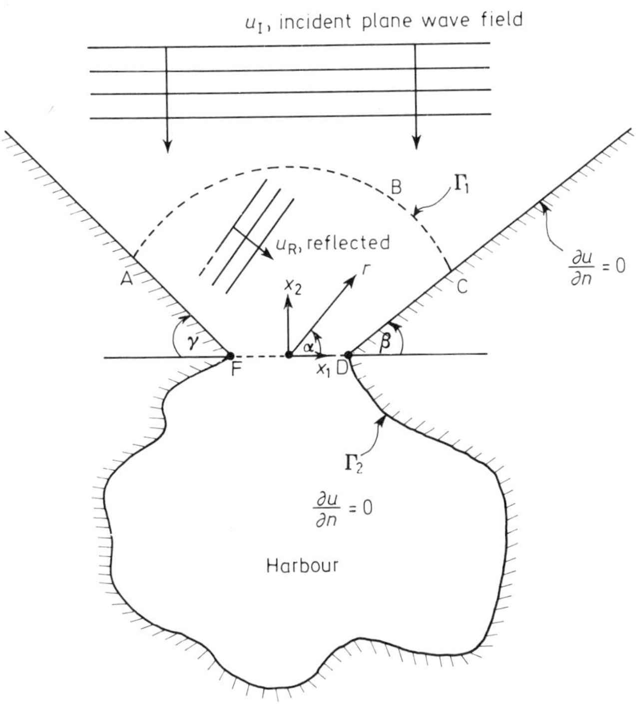

Figure 7.14 Definition sketch for harbour oscillation problem

(iii) the scattered field u_S originating from inside the harbour. Hence the total wave field is

$$u = u_I + u_R + u_S \tag{c}$$

with the incident field given by

$$u_I = a \exp[-i\kappa r \cos(\theta - \alpha)] \tag{d}$$

We know that the scattered wave field u_S will be travelling away from the harbour mouth on Γ_2. This fact is represented by the Sommerfeld boundary condition which may be written

$$\lim_{r \to \infty} \left[\sqrt{r}\left(\frac{\partial u_S}{\partial r} + i\kappa u_S\right)\right] = 0 \tag{e}$$

or approximately

$$\frac{\partial u_S}{\partial r} + i\kappa u_S = 0 \quad \text{on } \Gamma \tag{f}$$

using (f) we may write for the total field u

$$\frac{\partial u}{\partial r} + i\kappa u = \frac{\partial}{\partial r}(u_I + u_R) + i\kappa(u_I + u_R) = f \tag{g}$$

Note that the right-hand side is a known function f of the horizontal coordinates, the angles determining the local geometry and the angle of incidence of the wave field u_I.

We may now incorporate this boundary condition into our variational formulation to obtain

$$\int_\Omega \left[\frac{\partial}{\partial x}\left(h\frac{\partial u}{\partial x}\right) + \frac{\partial}{\partial y}\left(h\frac{\partial u}{\partial y}\right) + \kappa^2 hu\right] v \, d\Omega$$

$$= \int_{\Gamma_2} \left(h\frac{\partial u}{\partial n} - q\right) v \, d\Gamma + \int_{\Gamma_1} \left(h\frac{\partial u}{\partial n} + i\kappa u - f\right) v \, d\Gamma \tag{h}$$

which after integration by parts gives

$$\int_\Omega \left(h\frac{\partial u}{\partial x}\frac{\partial v}{\partial x} + h\frac{\partial u}{\partial y}\frac{\partial v}{\partial y} - \kappa^2 huv\right) d\Omega$$

$$= \int_{\Gamma_2} qv \, d\Gamma + \int_{\Gamma_1} (hf - i\kappa hu) v \, d\Gamma \tag{i}$$

which is the starting expression for our finite element formulation incorporating the boundary condition (f) on Γ.

Numerical results with, and without, the use of the radiation condition were obtained for the case of the old Duncan Basin within Table Bay in South Africa.

To simulate deterministically the effect of waves entering the harbour we discretised it into finite elements (Figure 7.15). Along the contour ABCDEF we applied the usual zero normal flux condition $\partial u/\partial n = 0$. On the harbour entrance AB we applied the radiation condition for an incident plane wave of unit amplitude entering the harbour in the negative y direction, for the moment ignoring reflections from the neighbouring coastline. The resulting elevations at node 70 for unit incident waves of differing frequencies are plotted in Figure 7.16. The first peaks occur at the same frequencies as those predicted by the harmonic analysis given in Brebbia and Adey.[9] For instance the first peak represents the first significant period of about 11·45 minutes which is the mode corresponding to water flowing in and out of the basin and is called the 'pumping mode'. The second peak also corresponds to a simple motion, the so-called 'sloshing mode'. The results were plotted up to $\omega = 0{\cdot}4\ \text{rad s}^{-1}$ which is the lower bound of the wind generated part of the wave spectrum.

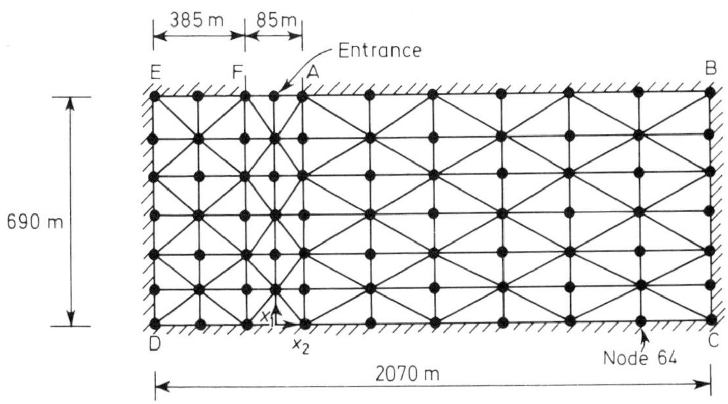

Figure 7.15 Mesh for Duncan Basin

It is interesting to note that after a series of mainly longitudinal modes, the transverse modes start to play a more significant part. The effect of these modes is considerable, especially from the probabilistic response point of view. The application of the radiation boundary condition on the harbour mouth adequately represents the

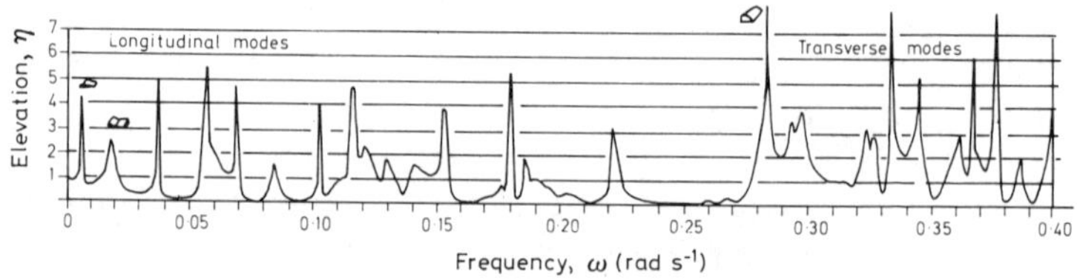

Figure 7.16 Maximum surface elevation at node 70 with radiation

input of wave energy into the system as well as energy passing *out* of the harbour. A more extensive discussion of this example can be found in Walker and Brebbia.[8]

7.4 INFINITE ELEMENTS

A technique that has recently received considerable attention is the development of finite elements extending to infinity. The technique resembles the approximate boundary element solutions developed in Section 7.3, but it is based on a completely different approach. It is due to Ungless[1] and has more recently been applied by Bettess.[2]

Here we start with the statement.

$$\int_{\Omega} (\nabla^2 u)\delta u \, d\Omega = \int_{\Gamma} (q - \bar{q})\delta u \, d\Gamma \tag{7.17}$$

used in finite elements and integrate by parts only once, to obtain

$$\int_{\Omega} \frac{\partial u}{\partial x_k} \frac{\partial \delta u}{\partial x_k} \, d\Omega = \int_{\Gamma} \bar{q}\delta u \, d\Gamma \tag{7.18}$$

We now look for decaying functions for u and δu. Consider for instance the case of the interface shown in Figure 7.17. The dashed lines represent zones to the interface emanating from the corner of the triangular element (which for simplicity is assumed to be of straight sides). The ζ coordinate emanating from the centre of the element can be used as a reference axis.

The function for the potential may be postulated as,

$$u(\xi_i, \zeta) = \bar{u}(\xi_i) f(\zeta) \tag{7.19}$$

$\bar{u}(\xi_i)$ are the usual two-dimensional functions such that,

$$\bar{u}(\xi_i) = \mathbf{\Phi}^{\mathrm{T}}(\xi_i)\mathbf{u}^{\mathrm{n}} \tag{7.20}$$

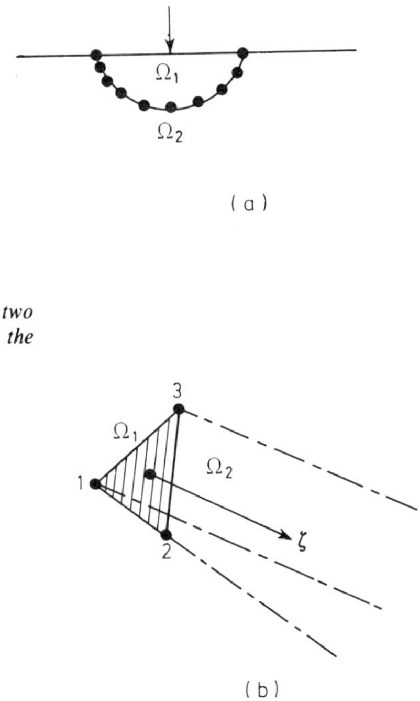

Figure 7.17 Infinite elements: (a) *the two regions;* (b) *triangular element on the interface*

where ξ_i are for instance the homogeneous coordinates on the triangular element. $\phi(\xi_i)$ are the interpolation functions and $\mathbf{u}^n$ are the nodal unknowns. The $f(\zeta)$ function is a decay function that reduces the magnitude of u as ζ increases. It is unity at $\zeta = 0$, and tends to zero at infinity. The function proposed by Ungless was simply,

$$f(\zeta) = \frac{1}{1+\zeta/L} \tag{7.21}$$

We can now substitute into equation (7.17) and work as in finite elements. The integration is carried out over the *infinite* domain of the elements and resort must be made to numerical integration.

Instead of function (7.21) Bettess[2] has used exponential decay. His shape functions in the ζ directions are defined as Lagrange polynomials multiplied by exponential decay terms. In this case at least a set of internal nodes is needed because of the polynomials, the simplest of which is the linear function with two nodal values.

The form of the solution will depend on the coefficients taken for the exponential decay, i.e. on the L of the following equation:

$$f(\zeta) = F(\zeta)\,e^{-\zeta/L} \tag{7.22}$$

The advantage of infinite elements is that they can be used for a series of problems for which the fundamental solution required for boundary elements may be difficult to determine. The method does

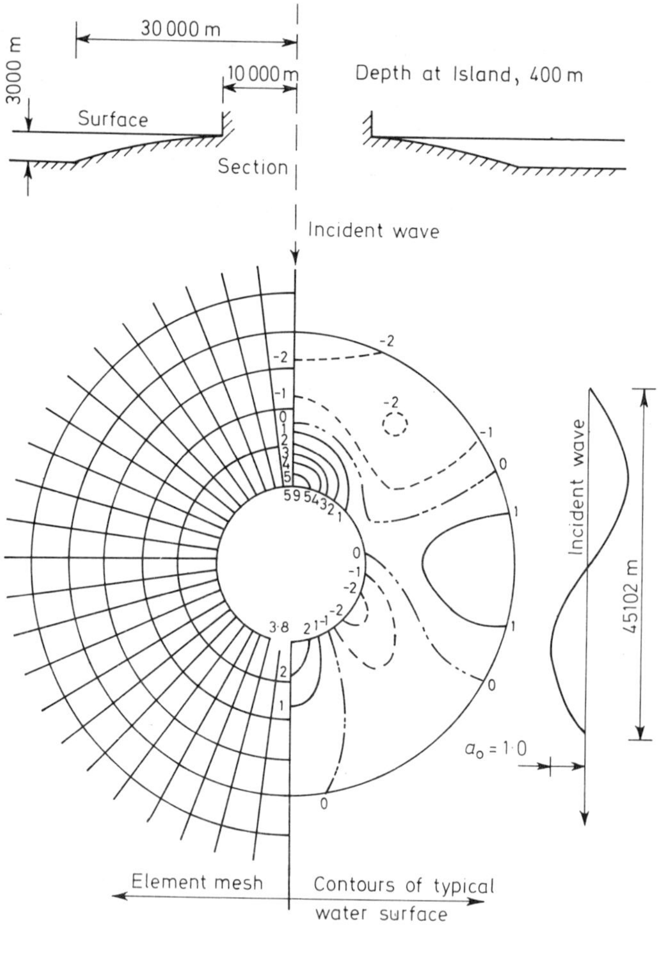

Figure 7.18 Wave diffraction by parabolic shoal and cylinder

not give an accurate indication of the behaviour of the system towards infinity but the effect of the far region on the domain of interest is introduced (see Bettess[2]).

Example 7.4[10] *Diffraction and refraction of waves by a parabolic shoal, surmounted by a cylindrical island*

Figure 7.18 shows the geometry of the problem and the element mesh used. The finite element mesh is enclosed by a number of infinite elements, the shape function in the radial direction is given by

$$P(r)\,\mathrm{e}^{-r/L}\,\mathrm{e}^{-\mathrm{i}\kappa r} \qquad \text{(a)}$$

(for time dependence $\mathrm{e}^{\mathrm{i}\omega t}$). Here $P(r)$ is a polynomial in r and L is the so-called decay length, and κ is the wavenumber corresponding to the frequency ω. This shape function satisfies the Sommerfeld radiation condition.

L was chosen so that near the island the decay $\mathrm{e}^{-3/L}$ roughly matched the decay of the first term of the general series solution, $H_0^{(2)}(\kappa r)$, where $H_0^{(2)}$ is the Hankel function of the zeroth order of the second kind. See Bettess and Zienkiewicz[10] for a fuller explanation of the above theory.

Figure 7.19, from Bettess and Zienkiewicz[10], shows a comparison of the elevations at the island itself compared with the analytic solution given in Homas[11] and Vastano and Reid.[12] The agreement is very good especially at moderate wavelengths.

The weakness in this method is that to determine the parameter L *some* knowledge of the exact solution is required before it can be used. However, the method is very efficient if this information is available.

7.5 COMBINATION OF FINITE AND BOUNDARY ELEMENTS

The idea of combining both techniques can be attributed to Wexler[13], who started to use integral equation solutions to represent the unbounded field problem early in the 1970s, the advantage being that this allowed for the use of appropriate conditions to represent the infinite domain. The integral equation technique is also of interest when regions of high stress or potential gradients exist, but finite elements are adequate for other parts of a body and may be simpler to

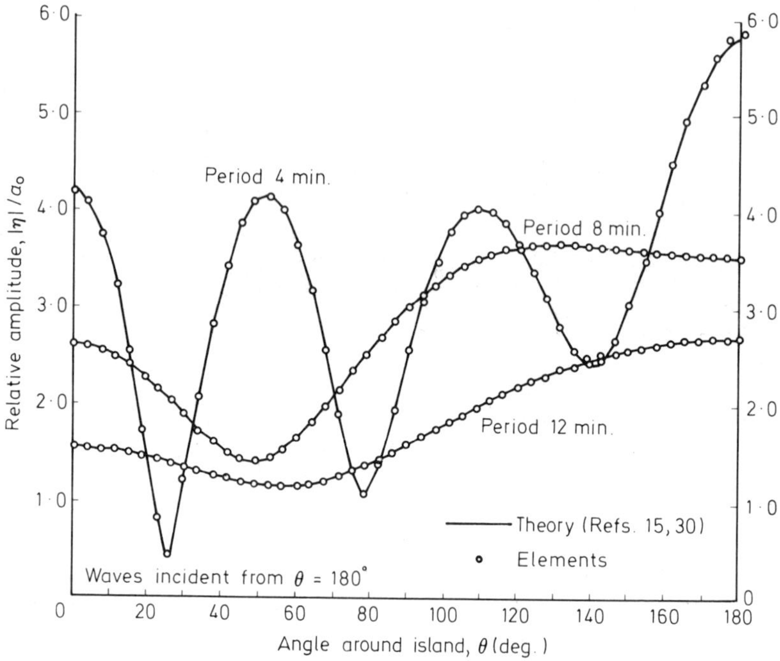

Figure 7.19 Relative amplitudes on cylinder

use in cases such as layered continua, anisotropic and non-linear materials. Hence it is important for the analyst to be able to represent a body using finite or boundary element techniques, depending on the particular geometry, boundary conditions, etc.

The first combination of the two methods for elastostatics appears to be by Osias[14], although for wave propagation problems, the method was used by Mei[7] in 1975 who explained the method of combining both solutions using variational techniques.

Lachat[3], following the work of the Southampton University group, developed a technique which allowed for variations of displacements and tractions along parts of the boundary surface or elements. The idea of using interpolation functions to define the variables along these elements allows for the combination of finite and boundary elements without any loss of continuity. In addition the work at Southampton[15] concentrated on the common basis of the different methods[16] and the equivalence of the direct and indirect boundary element techniques.

In this section the combination of boundary and finite elements is attempted for two-dimensional elastostatic problems and two different approaches are discussed here.[17] The one that appears more interesting is that for which the boundary element region is treated as a finite element, and can thus be easily incorporated into existing finite element computer packages (see Zienkiewicz, Kelly and Bettess[18]).

The technique is applied to a series of examples to illustrate how the combination can be done and how accurate the results are.

The starting expression for a finite element elasticity problem is,

$$\int_{\Omega} b_k u_k^* \, d\Omega - \int_{\Omega} \sigma_{jk} \varepsilon_{jk}^* \, d\Omega = - \int_{\Gamma_2} \bar{p}_k u_k^* \, d\Gamma \tag{7.31}$$

where u_k^* are the virtual displacements.

The one for boundary elements is,

$$\int_{\Omega} b_k u_k^* \, d\Omega + \int_{\Omega} \sigma_{jk,\,j}^* u_k \, d\Omega$$
$$= - \int_{\Gamma_2} \bar{p}_k u_k^* \, d\Gamma - \int_{\Gamma_1} p_k u_k^* \, d\Gamma + \int_{\Gamma_1} \bar{u}_k p_k^* \, d\Gamma + \int_{\Gamma_2} u_k p_k^* \, d\Gamma \tag{7.32}$$

where u_k^* represents the fundamental solution.

The standard finite element system can be written as

$$\mathbf{KU} = \mathbf{F} + \mathbf{D} \tag{7.33}$$

where $\mathbf{K}$ is the stiffness matrix for the system, $\mathbf{F}$ the equivalent nodal force vector and $\mathbf{D}$ the vector due to body force.

The vector $\mathbf{F}$ is obtained by weighting the applied tractions $\mathbf{p}_k$ on Γ_2 by the interpolation functions used for the displacements, i.e.

$$\mathbf{u}^{*\mathrm{n,T}} \mathbf{F} = \sum \left(\int_{\Gamma_2} p_k u_k^* \, d\Gamma \right)$$

where $\mathbf{u}^{*\mathrm{n,T}}$ is the vector of nodal virtual displacements for the complete system. The summation applies over the element sides on the boundary. Under these conditions we can write $\mathbf{F}$ as

$$\mathbf{F} = \mathbf{MP} \tag{7.34}$$

where $\mathbf{P}$ is a vector of nodal tractions and where $\mathbf{M}$ is a matrix due to the weighting of the boundary tractions by the interpolation functions for the displacements. The exact nature of $\mathbf{M}$ can be simply explained by considering the term

$$\int_{\Gamma_2} p_k u_k^* \, \mathrm{d}\Gamma \tag{7.35}$$

in equation (7.31), which is the boundary tractions weighted by the virtual displacements, or may be interpreted as the external energy expended due to a virtual displacement u^*. For a particular element, 'l', $\mathbf{p}_k^{\mathrm{T}}$ is a row vector containing the nodal value of the tractions and $\mathbf{u}_k^*$ a column vector containing the arbitrary virtual displacements u^*. (7.35) may now be written as

$$\int_{\Gamma_l} u_k^{*,\mathrm{T}} p_k \, \mathrm{d}\Gamma \tag{7.36}$$

Assuming the interpolation functions for u and p as

$$\mathbf{u}_k^* = \mathbf{\Phi}\, \mathbf{u}_k^{*,\mathrm{n}} \qquad \mathbf{p}_k = \mathbf{\Psi}\, \mathbf{p}^n \tag{7.37}$$

The above term reduces to

$$u^{*\mathrm{n},\mathrm{T}} \left(\int_{\Gamma_l} \mathbf{\Phi}\, \mathbf{\Psi}^{\mathrm{T}} \, \mathrm{d}\Gamma \right) \mathbf{p}^n \tag{7.38}$$

The matrix $\mathbf{M}$ may then be expressed as defined in equation (7.33). Hence we can write (7.32) as

$$\mathbf{K}\mathbf{U} = \mathbf{M}\mathbf{P} + \mathbf{D}$$

which is similar form to the boundary element equation, resulting from (7.32)

$$\mathbf{H}\mathbf{U} = \mathbf{G}\mathbf{P} + \mathbf{B} \tag{7.39}$$

Consider a problem consisting of two domains Ω^1, Ω^2 joined by an interface Γ^{I}, and which makes use of a finite element formulation in Ω^2 and a boundary element formulation in Ω^1 (Figure 7.20). In order to join the two parts we apply compatibility and equilibrium conditions along the interface Γ^{I}, i.e.

$$\mathbf{U}_{\mathrm{I}}^1 = \mathbf{U}_{\mathrm{I}}^2 \qquad \mathbf{P}_{\mathrm{I}}^1 + \mathbf{P}_{\mathrm{I}}^2 = 0 \tag{7.40}$$

where $\mathbf{U}_{\mathrm{I}}^l$, $\mathbf{P}_{\mathrm{I}}^l$ refer to the displacements and tractions on the interface Γ_{I} for the region l($l = 1, 2$ for a two-dimensional problem).

We now have two alternatives as to how to approach the problem. We may develop the boundary element region Ω^1 as an *equivalent*

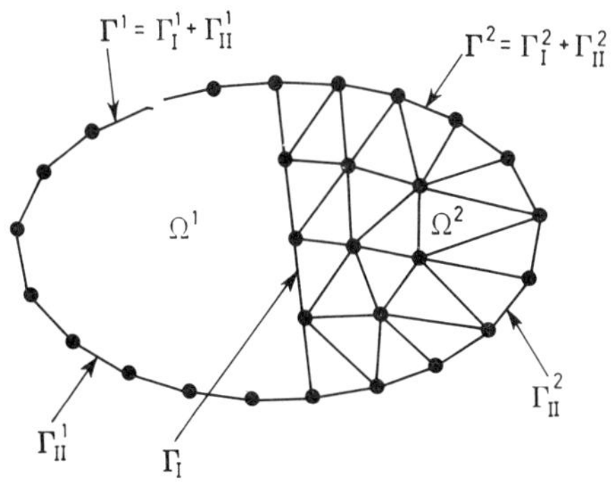

Figure 7.20 Domain divided into finite elements and boundary elements

finite element, assemble the effective stiffness matrix with those of the finite elements of region Ω^2 and solve the overall system as a stiffness problem. Alternatively we can consider Ω^1 and Ω^2 as if they were both *boundary element* regions.

Using the first approach, we can transform equation (7.39) by inverting $\mathbf{G}$, such that

$$\mathbf{G}^{-1}(\mathbf{H}\mathbf{U}-\mathbf{B})=\mathbf{P} \tag{7.41}$$

and premultiply by the matrix $\mathbf{M}$ defined by equation (7.34), giving

$$(\mathbf{M}\mathbf{G}^{-1}\mathbf{H})\mathbf{U}-(\mathbf{M}\mathbf{G}^{-1}\mathbf{B})=\mathbf{M}\mathbf{P} \tag{7.42}$$

We can now define

$$\mathbf{K}'=\mathbf{M}\mathbf{G}^{-1}\mathbf{H} \quad \mathbf{D}'=\mathbf{M}\mathbf{G}^{-1}\mathbf{B} \quad \mathbf{F}'=\mathbf{M}\mathbf{P} \tag{7.43}$$

Hence equation (7.42) has the following finite element form:

$$\mathbf{K}'\mathbf{U}=\mathbf{F}'+\mathbf{D}' \tag{7.44}$$

The main discrepancy which arises with the above formulation is the fact that the matrix $\mathbf{K}'$ is generally asymmetric although from first principles a stiffness matrix should be symmetric. The asymmetry arises due to the approximations involved in the discretisation process and the choice of the assumed solution. The matrix can be made symmetric by minimising the square of the errors in the non-symmetric off-diagonal terms.

The error for a coefficient k_{ij} can be written as the difference between k'_{ij} and k'_{ji} and the still unknown coefficient k_{ij}, which is symmetric, i.e.

$$\varepsilon_{ij} = \tfrac{1}{2}[(k_{ij} - k'_{ij}) + (k_{ij} - k'_{ji})] \tag{7.45}$$

The square of this error is now minimised:

$$\frac{\partial}{\partial k_{ij}}(\varepsilon_{ij}^2) = 2k_{ij} - k'_{ji} - k'_{ij} = 0 \tag{7.46}$$

Hence the new symmetric coefficients are:

$$k_{ij} = \tfrac{1}{2}(k'_{ij} + k'_{ji}) \tag{7.47}$$

The equivalent finite element type matrices of equation (7.44) may now be assembled with the matrices for the elements of region 2 to form a global system of equations. Assembling along the nodes of interface I ensures condition (7.40) and as the matrix **M** transforms the tractions **P** into equivalent consistent nodal loads **F**, by having zeros in the vector **F** corresponding to the nodes u_{I}, condition (7.41) is satisfied.

Using the second approach, mentioned above, for combining the two methods, we can consider region 2 as a boundary element type region.

For region 1 we can write,

$$[\mathbf{H}^1\ \mathbf{H}_{\mathrm{I}}^1]\begin{Bmatrix}\mathbf{U}^1\\ \mathbf{U}_{\mathrm{I}}\end{Bmatrix} = [\mathbf{G}^1\ \mathbf{G}_{\mathrm{I}}^1]\begin{Bmatrix}\mathbf{P}^1\\ \mathbf{P}_{\mathrm{I}}\end{Bmatrix} + \mathbf{B}^1 \tag{7.48}$$

and for region 2,

$$[\mathbf{K}^2\ \mathbf{K}_{\mathrm{I}}^2]\begin{Bmatrix}\mathbf{U}^2\\ \mathbf{U}_{\mathrm{I}}^2\end{Bmatrix} = [\mathbf{M}^2\ \mathbf{M}_{\mathrm{I}}^2]\begin{Bmatrix}\mathbf{P}^2\\ \mathbf{P}_{\mathrm{I}}^2\end{Bmatrix} + \mathbf{D}^2 \tag{7.49}$$

By writing $\mathbf{P}_{\mathrm{I}} = \mathbf{P}_{\mathrm{I}}^1 = -\mathbf{P}_{\mathrm{I}}^2$ and $\mathbf{U}_{\mathrm{I}} = \mathbf{U}_{\mathrm{I}}^1 = \mathbf{U}_{\mathrm{I}}^2$ we automatically satisfy conditions (7.40) and (7.41), and equations (7.48) and (7.49) can be rearranged as follows,

$$[\mathbf{H}^1\ \mathbf{H}_{\mathrm{I}}^1\ -\mathbf{G}_{\mathrm{I}}^1]\begin{Bmatrix}\mathbf{U}^1\\ \mathbf{U}_{\mathrm{I}}\\ \mathbf{P}_{\mathrm{I}}\end{Bmatrix} = [\mathbf{G}^1]\{\mathbf{P}^1\} + \mathbf{B} \tag{7.50}$$

and

$$[\mathbf{K}^2\ \mathbf{K}_I^2\ \mathbf{M}_I^2]\begin{Bmatrix}\mathbf{U}^2\\ \mathbf{U}_I\\ \mathbf{P}_I\end{Bmatrix} = [\mathbf{M}^2]\{\mathbf{P}^2\} + \mathbf{D}^2 \tag{7.51}$$

Writing these two equations together, as a single matrix equation, we have,

$$\begin{bmatrix}\mathbf{H}^1 & \mathbf{H}_I^1 & -\mathbf{G}_I^1 & \mathbf{0}\\ & & & \\ \underbrace{\mathbf{0}}_{\text{region 1}} & \underbrace{\mathbf{K}_I^2 \quad \mathbf{M}_I^2}_{\text{interface}} & & \underbrace{\mathbf{K}^2}_{\text{region 2}}\end{bmatrix}\begin{Bmatrix}\mathbf{U}^1\\ \mathbf{U}_I\\ \mathbf{P}_I\\ \mathbf{U}^2\end{Bmatrix} = \begin{bmatrix}\mathbf{G}^1 & \mathbf{0}\\ & \\ \mathbf{0} & \mathbf{M}^2\end{bmatrix}\begin{Bmatrix}\mathbf{P}^1\\ \\ \mathbf{P}^2\end{Bmatrix} + \begin{Bmatrix}\mathbf{B}^1\\ \\ \mathbf{D}^1\end{Bmatrix} \tag{7.52}$$

Notice that on the boundary of the finite element region Ω^2, only the displacements on Γ_1 have to be prescribed, whilst on the boundary of Ω^2 we prescribe the displacements or tractions and consequently need to reorder the equations.

The advantage of the second approach is that it does not require an inversion.

REFERENCES

1. Ungless, R. F., 'An Infinite Finite Element', *M.Sc. Thesis*, Department of Civil Engineering, University of British Columbia, Canada (1973)
2. Bettess, P., 'Infinite Elements', *Int. J. Num. Meth. Engng*, **11**, 53–64 (1977)
3. Lachat, J. C., 'A Further Development of the Boundary Integral Technique for Elastostatics', *Ph.D. Thesis*, University of Southampton (1975)
4. Brebbia, C. A. and Walker, S., 'Simplified Boundary Elements for Radiation Problems', *Res. Note, Appl. Math. Modelling*, **2**, No. 2, June (1978)
5. Chen, H. S. and Mei, C. C., 'Oscillations and Wave Forces in a Man Made Harbor in the Open Sea', *Proc. 10th Symp. on Naval Hydrodynamics, London, 1974*, pp. 573–94.
6. Shaw, R. P. and Falby, W., 'FEBIE—A Combined Finite Element-Boundary Integral Equation Method', *Proc. 1st Int. Symp. on Innovative Numerical Analysis in Applied Engineering Science, Versailles, France, May* 23–27, 1977
7. Mei, C., 'Oscillations and Wave Forces in a Man-Made Harbour in the Open Sea', *Proc. 10th Symp. on Naval Hydrodynamics, London, 1974* (1975)
8. Walker, S. and Brebbia, C. A., 'Harbour Resonance Problems using Finite Elements', *Adv. Water Resources*, **1**, No. 3, June (1978)

9. Brebbia, C. A. and Adey, R. A., 'Circulation Problems', *Proc. Finite Element Symp., Atlas Computer Laboratory, Didcot, UK March* 26–28, 1974
10. Bettess, P. and Zienkiewicz, O. C., 'Diffraction and Refraction of Surface Waves using Finite and Infinite Elements', *Int. J. Num. Meth. Engng,* **11,** 1271–40 (1977)
11. Hommas, S., 'On the Behaviour of Seismic Sea Waves round Circular Island', *Geophys. Mag.,* **21,** 199–208 (1950)
12. Vastano, A. C. and Reid, R. O., 'Tsunami Response for Islands: Verificiation of a Numerical Procedure', *J. Marine Res.,* **25,** 120–39 (1967)
13. McDonald, B. H. and Wexler, A., 'Finite element solution of Unbounded Field Problems', *IEEE Trans. Microwave Theory Techniques* **MIT–20,** No. 12, December (1972)
14. Osias, J. R., Wilson, R. B. and Seitelman, L. A., 'Combined Boundary Integral Equation Finite Element Analysis of Solids', *Proc. 1st Symp. on Innovative Numerical Analysis in Applied Engineering Science, Versailles, France, May* 23–27, 1977
15. Brebbia, C. A., 'Approximate Methods in Mathematical Modelling', *Proc. 1st Int. Conf. on Applied Mathematical Modelling, St. Louis, Missouri, August,* 1977
16. Brebbia, C. A. and Butterfield, R., 'The Formal Equivalence of the Direct and Indirect Boundary Element Methods', *Appl. Math. Modelling,* **2,** No. 1, March (1978)
17. Brebbia, C. A. and Georgiou, P., 'Combination of Finite and Boundary Elements for Elastostatics', to be published
18. Zienkiewicz, O. C., Kelly, D. W. and Bettess, P., 'The Coupling of the Finite Element Method and Boundary Solution Procedures', *Int. J. Num. Meth. Engng,* **11,** 355–76 (1977)

Index